Live Wire

Live Wire

Women and Brotherhood in the Electrical Industry

FRANCINE A. MOCCIO

TEMPLE UNIVERSITY PRESS
Philadelphia

Temple University Press
1601 North Broad Street
Philadelphia, PA 19122
www.temple.edu/tempress

Copyright © 2009 by Temple University
All rights reserved
Published 2009
Printed in the United States of America

♾ The paper used in this publication meets the requirements of the
American National Standard for Information Sciences—Permanence of
Paper for Printed Library Materials, ANSI Z39.48-1992

Library of Congress Cataloging-in-Publication Data

Moccio, Francine A., 1951–
    Live Wire : women and brotherhood in the electrical industry / Francine
A. Moccio.
        p. cm.
    Includes bibliographical references and index.
    ISBN 978-1-59213-737-4 (cloth : alk. paper)   1. Electric industry
workers—Labor unions—United States.   2. Building trades—
Employees—Labor unions—United States.   3. Sex discrimination against
women—United States.   I. Title.
    HD6515.E32U66 2009
    331.4'862130973–dc22   2008047846

2   4   6   8   9   7   5   3   1

You know the green neon lights we use on St. Patrick's Day or other holidays to honor the Irish, the Italians, and the different ethnics? I daydream about setting aside one day to honor tradeswomen electricians who tolerate harassment and risk their lives to earn a living and clear a path for all women. Pink neon is my choice—pink neon right there on top of the Empire State Building.

—JENNY, CONSTRUCTION ELECTRICIAN

Contents

Acknowledgments

This book grew out of my experience as a professor of labor studies in a unique program at the State University of New York (SUNY). I taught a course on class, race, and gender relations to blue-collar electrical apprentices enrolled in a college credit labor studies program. Since then, I have been a labor educator and working-women's rights researcher and advocate at Cornell University. The experience of teaching in the SUNY program sensitized me in my subsequent career to the complexities of understanding the challenges of addressing class, sex, and racial solidarity in the contemporary labor movement. The book is informed throughout by my close association for more than twenty years with craft unions in the building trades and with tradeswomen in New York, and influenced by the growing body of literature on sex segregation in the skilled trades.

A SUNY Faculty Research Grant, for which I am grateful, and the Elinor Goldmark Black Fellowship from the Graduate Faculty, New School for Social Research, supported part of this study. An evaluation of the women's progress in the New York City skilled building trades from 1978 to 2008 was in part the result of a financial settlement from an early sex-discrimination case litigated by Judith P. Vladeck, a renowned labor and women's rights attorney and close mentor and friend of the Institute for Women and Work (IWW) at Cornell University's School of Industrial and Labor Relations. The case, brought by Pamela Martens who sued Smith Barney, was chronicled by the popular book, *Tales from the Boom-Boom Room: Women vs. Wall Street.* Vladeck won a settlement that created a $6-million executive training program for women at Chase Manhattan Bank, now part of JP Morgan Chase & Co., and launched educational courses and research on women's status in nontraditional careers at Cornell's IWW.[1]

Among Vladeck's other prominent cases was her 1983 victory on behalf of female employees at Western Electric. At the time, it was the largest equal-pay case decided in favor of female plaintiffs, her firm said in a statement.

A book project involves efforts on the part of many people. I am very grateful first and foremost to the women and men electricians who gave so generously of their time for interviews during lunch breaks, in their homes, and on the job sites after work, especially Evan Ruderman, a pioneer woman electrician and AIDS activist who inspired a generation of women's rights advocates. Evan's sharp and provocative thinking were critical in shaping my insights into the industry.[2]

Enormous gratitude also goes to other veteran pioneer electricians: Joi Beard, Melinda Hernandez, Laura Kelber, Cynthia Long, Beth Schulman, and Jackie Simpson, as well as to the generations of women electricians who followed in their footsteps. Their moral courage to stand and fight sexism in the industry is nothing less than heroic. Many thanks are also owed to the following tradespeople and scholars who helped at various stages of the research: historian Mary Ann Clawson for help on an early revision; Cheryl Farrell, Earline Fisher, Dan Georgakas, James Haughton, Rocco Rorandelli, and Veronica Rose; feminist scholars Martha Fineman, Miriam Frank, and Heidi Hartmann. My co-conspirators in women's rights deserve special mention: journalist and author Jane LaTour, and legal scholar Barbara Phillips. I am ultimately responsible for any remaining errors in the book.

I owe a great intellectual debt to the many scholars whose work is cited here, but some merit special mention. My deepest debt is to my mentors at the Graduate Faculty, New School for Social Research, in Manhattan. William Roseberry, professor of anthropology at New York University and the Graduate Faculty, New School, encouraged this research from the beginning, especially the unconventional idea that doing fieldwork in the New York construction industry would be as "exotic" as being airlifted into Papua, New Guinea. Roseberry suggested interweaving historical and contemporary ethnography with personal narrative to explain the gender dynamics of the industry without having it killed by science.[3] Rayna Rapp, one of the foremost feminist anthropologists, provided important mentorship and advice on an earlier version of this research. Eric Hobsbawm, a visiting lecturer at the New School, first introduced me to the importance of studying fledgling grassroots networks that challenge formal power and authority. Hobsbawm's guest lectures on grassroots labor sects and nineteenth-century proto-trade-unionism of craft journeymen and apprentices prompted me to look more closely at Local 3 social clubs and rituals. The flowering of old and new symbolic forms in the electrical brotherhood are deeply intertwined with the ritual structuring of workers' personal lives and social collectivities, and with rites of passage of young men to citizenship and manhood. This amalgam of "invented traditions" in the union with its symbols, rituals, and logos represents an important form of present-day union solidarity.[4]

I owe a great debt to Stanley Diamond, a renowned anthropologist, mentor, and friend, for his guidance on an earlier version of this study. Diamond encouraged anthropologists to recognize that the most pertinent issues surrounding explanations of social relations emerge in the field. The New School encourages interdisciplinary research, and to this extent the Economics Department was a close ally of the Anthropology Department. David Gordon, professor of labor economics, merits special mention for his insightful guidance on the role of political economy in trade unions and for his help as adviser on my doctoral thesis.

I am deeply grateful to Betty Friedan, noted feminist and author of *The Feminine Mystique* (1963), for mentorship and advice on this book and related projects at Cornell's Institute for Women and Work. Friedan believed that narrow social science approaches have many shortcomings; as she stated, "That's not the way life is." In addition, at Cornell's ILR School, Professor Alice H. Cook's early article on the role of social clubs in the electrical brotherhood influenced my thinking about the role political clubs play within the union. Professor Maurice F. Neufeld's helpful review of an earlier version of this research and his groundbreaking article on the electrical brotherhood's complex administrative operations were invaluable to the book. My faculty colleagues at ILR and Cornell deserve special thanks for their ongoing encouragement and support of the book: Fran Blau, Ileen A. DeVault, Maria Figueroa, Lois Gray, Risa Lieberwitz, Peggy Sipser, and Amy Villarejo. Marty Wells and Kathy Roberts provided much-needed research support. I deeply appreciate their friendship, advice, and generosity.

For assistance with retrieving archival documents at the Wagner Labor Archives at New York University's Bobst Library, Debra Bernhardt, a labor scholar and activist, deserves special mention. Yalda M. Haery, my research assistant, provided valuable assistance in gathering secondary sources and documenting archival materials.

For their insightful editing of earlier drafts of the book and for their unwavering support and encouragement, much gratitude and thanks go to Priscilla Murolo, director of the Sarah Lawrence Graduate Program in Women's History, and A. B. Chitty, systems librarian at Queens College (CUNY), coauthors of *From the Folks Who Brought You the Weekend* (The New Press, 2002). In addition, Elizabeth Oguss, Rick Perlstein, Meredith Phillips, and Meg Fergusson provided some advice and copyediting on the book. Thanks also go to Mick Duffey-Gusinde, my editor at Temple University Press, who believed in the book and gave important and critical guidance for its completion.

Last, gratitude to my family is unquestionable, especially to my husband, Petros Anastasopoulos, Professor of Economics at Fairleigh Dickinson University in New Jersey, who gave generously of his time to assist in running and analyzing census data for the book and who enjoyed tagging along on some fascinating field trips to Bayberry Land, the union brotherhood's former recreational facility in Southampton, Long Island. In addition, I owe a great deal of thanks to my two older sons, Jason and Tony, who rode on front and back loaders, cherry pickers, and cranes and listened to endless stories about construction work. To the latest additions to our family, Alexandros and Angelos, who sacrificed going to the playground so that I could finish writing, and to Ligia Acosta, a kindred soul, I remain eternally grateful. All of my children are my undeserved bonuses. But it is my late father, Gennaro Carmino Moccio, an Italian crane operator and an active union brother and member of the International Union of Operating Engineers (IUOE), Local 15A, who is the major inspiration for this book. The pride he took in his craft, and the many conversations with his work buddies I was privy to while growing up in our Brooklyn apartment, helped me to shape invaluable and early insights into the experiences, intelligence, wit, and humor of construction workers. This ultimately shaped my research interests in the construction industry, its workers, and its unions.

Introduction: Getting Wired

Institutionalized discrimination based on race, class, and gender is a persistent theme in American history and an enduring trait of our culture.[1] Although significant progress has been made in integrating women in the professions, one of the last bastions of male dominance is the construction building trades, which is 98 percent male.[2]

The 2009 American Recovery and Reinvestment Act has allocated $789 billion for 3.5 million new jobs over the next year for a range of industries (90 percent of which are in the private sector) from infrastructural building work to energy renewal to health care. Nationally, $150 billion is set aside for infrastructural and building work, the largest allocation since the federal interstate highway system.

For states like New York, this allocation translates into the creation of over 215,000 new jobs. In New York City, the rebuilding of the Twin Towers site alone in lower Manhattan will generate over 26,000 new on-site construction jobs and approximately $3.5 billion in direct and indirect wages.

Today, the creation of "green" jobs for new construction will also increase employment opportunities in the construction sector. Preexisting buildings now require retrofitting for sustainability, energy efficiency, and weatherizing. These new jobs and conversion projects are federally subsidized. Accordingly, the awarding of procurement contracts requires eligible contractors and unionized apprenticeship programs to comply with federal and state regulations and statutes on nondiscrimination in recruiting, hiring, and training women and minority workers.

In bidding for these contracts, employers are obligated to adhere to governmental agencies' affirmative action plans by submitting goals and timetables that show a "good faith effort" on their part to fulfill these goals. Today, opportunities exist for no- and low-income women, as well as women in general, who are interested in on-the-job training with good pay. Nonetheless, history has shown that when the "gravy train" arrives, women and minorities do not automatically get on board.

A complex set of historical and contemporary factors fuel sex discrimination despite legislative reforms and decades of feminist advocacy. Whether the construction trades have universal appeal as a career pathway for women is not nearly as important as understanding why this severe exclusion of women persists; what role workers, unionists, and employers play in sex segregation; and how women have organized their identity for collective action in response.[3]

The present-day gender war between men and women in the building trades and the events leading up to it provide a microcosm of how networks of male solidarity, in the form of fraternalism and brotherhood, stand in the way of women. This "war" in the electrical industry and trade also illustrates the degree to which social movements for equality of the 1960s and 1970s and their decline in the 1980s influenced workers' attempts at sex and race desegregation of the labor market. Women's responses to this exclusion, divided along the lines of gender, race, sexual preference, and economic status, provide insights into the clash between brotherhood and the feminist movement.

When I first began this research (although I am embarrassed to count the years), I fully expected that inclusion in the building trades would inevitably evolve due to labor market forces such as the shortage of skilled labor resulting from the anticipated retirement of workers from the baby boom generation. However, I was intrigued by a very simple observation which led to the essential question in this book: What was all the resentment and fuss about from male building contractors, unionists, and workers about a handful of women in the trades? I intended to simply explain the hostile behavior of men toward women in the electrical trade as a function of their class status as the proverbial "aristocracy of labor." I quickly realized the story of women's struggle in the industry was much larger and the collective actions of these tradeswomen for dignity and a middle-class wage much more complex. The brotherhood was rife with contradictory values that are at once egalitarian and exclusionary, proud and submissive, militant and accommodationist. "If it was easy, old men and women could do it," a construction foreman tells me at the Jacob Javits Center in New York City.[4] "Women are for after work," a male ironworker explains while we drink in a Boston pub.[5] Curiously, men I interviewed in job crews in the construction industry drew upon what is now generally seen as outdated polarized views of gender relations in American society: "You put a woman in that crew, and the work just slows down," asserts a major electrical contractor from Philadelphia. "Women have small fingers and are much more suited to manufacturing," says my male plumber friend. "Anyway, a woman in my trade is just taking the place of a man who's trying to feed his family."

No, that is not a scene out of a 1950s movie. These remarks were made recently about women across the various building trades including women electricians. Why do working men in the trades, as well as their bosses and union representatives, still have what appear to be fossilized attitudes? Why are women are still excluded? This book attempts to answer these questions.

February 2001

Anniversary memorials at the former site of the World Trade Center Twin Towers in downtown Manhattan continue to mark the tragic events of September 11, 2001. But the ghost-like structures of the twin towers still loom against the New York City skyline and reverberate like phantom nerve endings on an amputee. Rewind back to a freezing morning in late February 2001. A cold wind blew off the Hudson River, white cumulus clouds surged past the skyscrapers of Wall Street, and the harsh wind blowing off the river bit into my face. I pulled my woolen hat over my ears and waited for my respondent to show up. For shelter, I huddled in a corner doorway of what were then the two tallest buildings in New York City—the World Trade Center.

Memories of my youthful experiences at the World Trade Center flooded through me. Suddenly it was 1967 and on a job site in lower Manhattan my father was climbing up a long snake-like ladder anchored in the bowels of the earth. He was arising from the newly dug foundation of the World Trade Center. He emerged in his dirty work clothes onto the midday sidewalk to spend his half-hour lunch with me, a high school junior, so he could "show me the job." I climbed gingerly back down with him into the pit from which he had just emerged. At the bottom, I stepped off the ladder and stood in a huge hole that resembled the impact site of a super-sized meteorite.[6] My father and I stood silent for a moment, hundreds of feet below the sidewalk. As one would imagine the vertical prowess of a basketball player by peering at his stunningly oversized shoes, I tried to imagine the building destined to rest in this immense void. My father turned to me and asked in mixed Italian and English, "So, I helped do this baby. Ti piace? [Do you like it?]"

Three years later, I again joined my father at the trade center building site, which was now about fifty stories high. This time, though, I did not come for a tour with my dad. Instead, I was there with hundreds of my peers for a demonstration of college students protesting the war in Vietnam.[7] My father was on the opposite side. The protest quickly became confrontational. Angered by the raucous shouts of student leaders, the construction workers in their shiny hard hats of red and yellow and blue raced toward the students, brandishing work tools and clubs. They chased the students through the streets of New York's financial district, attacking whomever they could. I saw blood spurting from defenseless protesters. The workers seemed to be driven by a force beyond their control. But in the decades to follow, this violence came to vividly symbolize much more: the split between American labor and youth and a rift that is only now slowly starting to close.

My father's opinions about the war, however, changed that very day. He viewed the attack on the students as an attack upon his daughter. He became increasingly sympathetic to the antiwar movement and increasingly critical of the U.S. role in Vietnam.

Thirty-one years later, on February 27, 2001, hundreds of employees streamed out of the huge Venetian cathedral-like doors of the Twin Towers. A virtual lunchtime ballet of secretaries, executives, financial analysts, and bankers emerged into the frigid afternoon. The sheer number, energy, and diversity of the people pouring onto the noonday sidewalks were quintessentially New York. Checking and rechecking my watch for accuracy, I wondered if my interviewee, a female electrician, a woman in a man's job in one of the most tenaciously gender-segregated and masculine-sex typed industries in America, was going to show up at all.

Then someone caught my eye. I idly watched a slim young man with a blonde ponytail striding toward me. Dressed in khakis and brown construction boots, this person drew closer—and it suddenly became clear that what had looked like a man in the distance was actually a woman. She wore a Local 3 hat and a lumberjack coat with the collar turned up. Her name was Jenny, and she had granted me the interview because she was leaving the industry and wanted to tell someone her story. Joining the throng of people on the sidewalk, we headed down the chaotic alleys of lunchtime Wall Street in search of a quiet place to talk.

Halfway down the block, she gripped my arm and pulled me up short. She pointed to one of the tallest skyscrapers in sight. "Do you see that pretty neon sign up there?" She asked. "I wired that. Every time I pass by with my family or friends, I show them that. I know I won't ever get that feeling from working in a day care center." She spoke in a heavy Southern drawl.

We ducked into a small café and my tape recorder began to roll. Only seven years before, Jenny, a single mother of two, had moved from Georgia to New York City, breaking a three-generation family pattern of relying on public assistance. Soon she was inducted into electrical apprenticeship.

On her way to an $85,000-plus per year job as a journeyman electrician, she ignored the girlie pinups and the wisecracks of the guys, keeping her eyes on the prize—a middle-class existence for herself and her children.

Life seemed full of promise until the day Jenny's foreman handed her a power drill and ordered her to work close to a high-rise elevator shaft under construction, even though he knew she was inexperienced with that tool. The tool's recoil jolted her back toward the empty shaft. Grabbing onto one of the steel beams, Jenny found herself precariously dangling fifty stories above the street.

There were other incidents; once she saw the sole female worker on an otherwise all-male crew tied to the upper end of the Brooklyn Bridge's vast expansion cable. Incidents like these made her fear for her physical safety as an electrician. Jenny hoped that once she passed the apprenticeship stage and won her journeyman's card, the sexual hazing and harassment by co-workers and supervisors would stop. It did not. Instead, her journeyman status provoked even greater hostility from the men. She had been "mooned" by men on ladders, "dyke-baited" because of her attempts to organize women in the trade, and accused of sleeping with the boss when she was promoted to foreman. Now, at the age of thirty-eight, Jenny was thinking of taking an entry-level job as a daycare worker with an annual salary of $35,000 in order to escape the emotional stress of being a female electrician.

Is Jenny's story an isolated case? Not at all. After many years of research and activism on behalf of women in the construction trades, I would say that it is typical. Again and again I have heard stories from women electricians, as well as from women in other trades, that end with their leaving the industry, usually for lower paid, less skilled, less prestigious work. It is no wonder that women pioneers in the trades never became settlers.

Employers and unionists promote the view that women like Jenny leave the construction trades because they "can't hack it." Conventional wisdom among contractors and unionists holds that "women don't like to do dirty work." Male coworkers say that women get special privileges or try to advance on the job by trading sex for favors. During periods of economic contraction, men justify the gross underrepresentation of women in the skilled trades with arguments like "men must support their families" and "a woman should not take the place of a guy." During periods of economic expansion, men still exclude women under the premise that women do not view construction as their life's work. One contractor stated: "What it boils down to is that women are simply not interested." Bust or boom, women workers do not get their share of jobs in construction.

Male contractors, union leaders, and workers alike express disdain and hostility to the women who try to work as electricians, plumbers, carpenters, heavy equipment operators, or steamfitters or in other skilled trades. Only 2 percent of people in the skilled trades are women.[8] Over the past one hundred years, craft unionists, especially electricians, have to some degree, either willingly or by court decree, inducted men of diverse ethnicities and nationalities into their trades. So what is all the fuss about a handful of women? I hope the ensuing chapters of this book will help identify the barriers women in general face, which are especially complex for women of color.

Women first entered electrical construction and other skilled trades as a result of feminist mobilization in the late 1970s. Efforts to narrow the gender wage gap included desegregating the well-paying blue-collar jobs that had been, for the most part, male preserves. The pioneer females who entered the industry found their networks of support dismantled within a decade by decreasing federal support, weakened enforcement of civil rights legislation, and waning social activism. Even though high technology jobs, such as engineering, get more public attention today, skilled trades people such as construction electricians are also in great demand. Not only is construction work expanding nationally, and related occupations such as computer maintenance and repair are predicted to grow over the next ten years, new areas such as alternative energy and sustainability technology are also emerging in building construction.[9] Similar to other industries, electrical construction is set for a major demographic shift among field workers and industry leaders. For example, the group of 55 and older electricians—those closest to retirement—grew from 33 percent to 39 percent over the past two years. The youngest age group in the industry is shrinking and experienced leaders will retire en masse over the next two decades.[10]

Because traditionally male jobs such as construction electrician require only a high school diploma and on-the-job training, they provide an avenue for economic mobility for low income and poor women—especially for single mothers, women of color who may have few options for higher education, and former welfare recipients.[11] Conventional wisdom today holds that there has been progress desegregating the sexes in the labor market. After all, some strides have been made in the fields of law, medicine, and, lately, even in male-dominated professions such as engineering. Few consider the opportunities that blue-collar work, such as the building trades, can provide for women despite the fact that in New York City alone, the construction industry overall generates approximately $26 billion in revenues annually and the contract value of construction nationally is over $600 billion. Yet stories like Jenny's are not widely known, and blue-collar women do not figure much in the consciousness of professional women who profess and promote feminism.

After three hours at the Wall Street café, Jenny and I finally put on our coats and headed back out into the cold. We did not want to part: we met as total strangers; now I knew something of Jenny's ten years of experience in the trade—the good, the bad, and the ugly, funny stories and frightening moments alike. Pausing in the street, she turned to me in appeal:

If I can only make men understand that I love this work. You know, it's like growing children—you work and tend to their needs, you feed them and you clothe them, then one day, they are human beings bringing you joy and love. Buildings are like that, too. You nurture them from infancy to adulthood, from the ground wire to the top of the antenna, working every floor, investing time, labor, and good materials. Then, wow! One day, the building's done. The foreman throws the light switch and it's wired to glow!

For Jenny, becoming an electrician meant more than just a high-paying job. Wiring buildings brought her intense pride and satisfaction. And she knew she was forging the way for others.

Sometimes I think about how women in this industry are still pioneers despite the fact that they arrived over twenty years ago. This may sound crazy, but you know the green neon lights we use on St. Patrick's Day or other holidays to honor the Irish, the Italians, and the different ethnics? I daydream about setting aside one day to honor tradeswomen electricians who tolerate harassment and risk their lives to earn a living and clear a path for all women. Pink neon is my choice—pink neon right there on top of the Empire State Building.

We said goodbye. I watched Jenny disappear in the evening rush hour crowd, just as I had watched her approaching earlier that day. Her construction boots and khaki pants stood out boldly against the Armani suits, the Gucci briefcases, and the high-heeled shoes of the white-collar workers now going home. As Jenny walked away, I looked up at the massive buildings she had, in her own words, "nurtured," and reflected on her strange analogy. Putting up a building and raising a child: such a traditionally feminine way, I thought, to describe such traditionally masculine work. I still hear from Jenny now and then. Her experiences and women electricians like her inform this book.

My training as an anthropologist influenced my interest in presenting a micro view of gender relations in construction within a macro context of larger societal opportunities and constraints. I looked for culture in the most common of places among electricians—in bowling alleys, offices, coffee shops, bars, apartments, and classrooms, as well as on work sites and city benches during lunchtime. Unlike other accounts of women in construction, I place the work site experience of women and their failure to get their proportionate share of jobs in the skilled building trades as electricians within the larger scope of preexisting institutional fraternal structures of brotherhood and the electrical industry.

I have listened carefully to male and female workers, union leaders, union dissidents, activists, and employers, and closely worked as a researcher and educator with craft unions across a broad scope of building trades and with one New York City electrical union in particular—its leaders, workers, fraternal clubs, and affiliated employers.

Personal Background

No doubt my childhood led me to a lifetime of interest in the labor movement. I grew up in a blue-collar Italian American neighborhood in Brooklyn. As a child, I had firsthand experience of the economic insecurity and gender polarity in families headed by male building tradesmen, which was what most of the student electricians I later taught expected to become.

The scene in our Williamsburg, Brooklyn, apartment was tense on Sunday nights as my father anxiously awaited "the call" for work the next day, and its location. Feeling lucky to get a work "ticket" at all, and fearful that a wrong turn en route to the job the next morning would result in a layoff, my father would always rehearse the best route to the job. It was our Sunday night ritual: we all piled into his blue 1958 Chrysler to accompany my father as he made his trial run. On these trips, I recall, I admired him as he pointed to some of the tallest and most beautifully ornate skyscrapers in New York City, repeatedly pointing out, "I built that."

I also remember that I cried as a small child when I first noticed his filthy fingernails at the end of the day. He could never get them entirely clean. My father used to hang his dirty work clothes, often smelling of diesel fuel, in a white metal cabinet in the hallway before my mother would allow him to enter our three-room apartment. I was in charge of taking these smelly work clothes every week to the Catholic Church-owned Christian Brothers Dry Cleaning shop on Humboldt Street—a special dry cleaner for laborers that had a certain opprobrium, even among my working-class neighbors in the small but tight-knit community of Williamsburg.

Paradoxically, despite the anxiety and insecurity that seasonal work such as construction brings to families like my own, the construction work itself—with its physical prowess, economic dignity, and high visibility—helps form a "king of the castle" male identity at home. When she married my father in 1946, my mother, like so many of her contemporaries, quit her job (as a theatrical costume seamstress for Broadway) to become a wife and homemaker. She was strictly devoted to taking care of her family, freeing my father from all household and child care responsibilities so he could work at his job and volunteer for union activities. No one challenged this arrangement.

As long as blue-collar craftsmen could provide enough to support their families, not even the ensuing women's movement, galvanized by Betty Friedan's groundbreaking book *The Feminist Mystique* in 1963, could alter gender relations in blue-collar families.[12] The craftsman's primary role as family breadwinner in America's post–World War II nuclear family, made possible by a family wage, helped sustain traditional gender relations at home.[13]

My intimate experience and understanding of working-class blue-collar values has always remained a source of conflict with my ideals as a practicing feminist. I still feel a great deal of empathy for my electrician students and their union, as well as construction workers generally, despite their stereotyped attitudes toward women. These blue-collar students work outside all day, sometimes in the freezing cold, sometimes in sweltering heat, and often under hazardous and unpleasant conditions.[14] They attend night classes and support their families. The union tries desperately to keep construction work unionized in a political climate and city where union-busting tactics of large corporations such as Disney as well as other, usually smaller, union-busting contractors[15]—organized and supported by the powerful political lobby the Associated Builders and Contractors Inc. (ABC)—are joined by a federal assault on longstanding labor legislation.[16] The economic interests of owners and contractors have prevailed in their efforts over the years to degrade the skill of the craft worker by replacing older craft tools with new time-saving devices and by the use of prefabricated materials on work sites.[17]

With regard to technology, electrician journeymen have been bombarded with technological innovations designed to streamline production and reduce labor costs, such as the introduction of fiber optics. Moreover, with the rising costs of construction, management is looking to cut labor costs. The introduction of electronic technologies such as digital record keeping allow management to better monitor work crews' performance and measure productivity of workers and straw bosses against project estimates. The changing scope of electrical construction work from its former emphasis on heavy manual labor to an increased focus on mental labor require forced applications of such new technologies as digitized voice, videotaping, and recording; communications/systems connectivity; and alternative energy and sustainability building technology. In general, electricians face many challenges over the coming decades to retain their autonomy and control over the craft and the labor supply to contractors while keeping pace with technological change requiring more sophisticated training and staving off attacks on prevailing wage laws.[18]

Management's drive to reduce spiraling construction costs also aims at hiring a cheaper workforce by circumventing unionized apprenticeship programs. Thus unions like the electrical brotherhood encounter a great deal of resistance from small but well organized contracting and subcontracting groups which lobby on behalf of the nonunion movement at the state and national policy levels.

Paradoxically, as traditional cultural definitions of masculinity and femininity are fading, polarized gender relations such as machismo and that "king of the castle" mentality still resonate among building tradesmen. But tradesmen are no longer "masters" as they once were at work and at home.[19] Let us step back into my electrical apprentice class: understandably, it was quite a challenge to discuss social movements for racial and especially gender equality with my electrical-apprentice students without causing them to reach for their soldering irons (the electrical version of a red-hot poker).[20]

Although my students bluntly revealed their hostility and bitterness toward the entry of women and minority men into "their" trade, I felt it important to deal honestly and openly with these attitudes, knowing that behind this politically incorrect thinking lurked the fear of unemployment and economic insecurity. One might say, "She took the place of my buddy—he couldn't get in because of her." Another might claim, "Women can get married and let their husbands take care of them." Most would agree: "If it wasn't for the government interfering in our business, all these minorities would not be able to take over."

The college-level course I taught in a union-sponsored bachelor's degree program for electrical apprentices included the history, culture, and economics of the electrical construction industry and their union; the impact of the civil rights movement of the 1960s and the feminist movement of the 1970s on the industry and trades. It also included the anatomy of the electrical brotherhood; that is, the structure and function of the fraternal social clubs in the union, the social and economic changes affecting change in the craft and the industry, the growth of nonunion firms, and the danger of deteriorating health and safety standards on job sites; as well as how race, gender, and ethnic divisions can divide the union membership and weaken solidarity. Explaining affirmative action and compliance laws as a way to right past injustices against women and minorities in the trades was not going to go over well with these working-class students. I decided instead to use the history and idioms of their own trade and union to show how workers—women and men, black and white—have been divided by employers throughout history. I discussed with them how keeping out blacks and women would ultimately yield to a divide and conquer strategy on the part of contractors, and eventually weaken the brotherhood. I tried to illustrate through readings, lectures, and discussion of discrimination court decisions how their own economic security was threatened by the possibility that nonunion contractors would hire women and minorities at below-union wages. This got their attention.

At about the same time, although not a skilled tradeswoman, I started becoming active in the union and as a professor of labor studies at the State University of New York, I was recruited by the Local and by tradeswomen to deliver special lectures on topics related to sex discrimination in employment and employer and union obligations under Title VII of the Civil Rights Act of 1991. I delivered these educational workshops to male unionists at the union's educational center in Flushing, Queens, and at Bayberry Land, a former union-sponsored summer retreat in the Hamptons for workers and their families. As an academic and labor educator, I worked then and continue to work with tradeswomen and their organizations to address their grievances.[21]

Setting, Research Method, and Relevant Literature

I was able to conduct most of my fieldwork in close proximity to where I grew up and lived, New York City, an important center for commercial building construction work. In order to illustrate how male solidarity, at first a mechanism to combat economic exploitation in the building trades by organizing strong union brotherhoods, in the twenty-first century results in exclusionary practices, I present a case study of the International Brotherhood of Electrical Workers (IBEW), Local Union 3, and the Electrical Contractors' Association (ECA), which are the most powerful brotherhood and contractors' association in the industry today. An employers' association, the ECA represents approximately fifty of the largest and oldest electrical contractors in New York City (and the country, for that matter). The ECA is comprised entirely of white men, many of whom are former Local 3 Division A journeymen (these electrician journeymen are union members who are among the highest-paid tradesmen in the country and in the construction industry[22]).

Local 3 is viewed as more politically powerful than the IBEW itself. The Local is structurally linked to the ECA through its powerful Joint Industry Board (JIB) and has tremendous political influence over other construction trade and craft unions. Currently, the union has approximately thirty-five thousand members, thirteen thousand of whom are in Division A, the elite construction division. Typical of the gender composition of the electrical industry, the remainder of the members is mostly minority women in electrical manufacturing and related divisions and occupations, including secretaries. In order to protect my respondents and their jobs, the true identities of some female and male respondents have been kept anonymous; in those cases I have substituted first name pseudonyms throughout the book. In other instances, the true names of respondents appear, with both first and last names.

Local 3 is characterized by its highly evolved structure of cross-class, cross-generational, and cross-cultural relations among workers, unionists, and employers. Internal stratification among electrical workers takes the form of different classes of workers: journeyman (highly skilled labor), apprentice (cheaper, quasi-indentured labor), and traveler (unionized electricians from other parts of the country).

The Local is organized into thirty divisions by industry, such as building construction, street lighting, electrical manufacturing, and clerical work.[23]

Other organizational elements include the Apprenticeship Program, the JIB, and eighteen strong fraternal ethnic associations known as "social clubs." These fraternal social clubs, common to many locals in the electrical workers' brotherhood but most highly evolved in Local 3, exist within and alongside the Local. Like their counterparts in the late nineteenth century such as the Knights of Pythias and the Masonic societies, the social clubs allow the brotherhood to efficiently administer union benefits and services to a large membership and serve as a training ground for prospective male union leadership. The clubs provide union members with opportunities to bond as brothers around their own identities of race, ethnicity, religion, and occupation, and to access commercial information vital to keeping jobs unionized.[24] Symbolic rituals lace every aspect of club functions and play an influential role in the brotherhood by infusing a sense of belonging on the part of its members. Every breakfast forum and social event incorporates symbols of male leadership, for example, gatherings are prefaced with a salute from various Boy Scout troops. Furthermore, social events such as the Ladies' Honor Scroll Nights, the Ladies' Nights dinner-dances, and the raffling of jewels and fur coats place wives of male electricians on a pedestal.

But we must not be fooled. Clubs, apprenticeship programs, sports clubs, fraternities, the Electchester housing complex, the JIB bowling alley and child care center, the family compound of Camp Integrity for members' wives and children, examined in the context of how some of their union husbands and male electrician spouses mistreat women on the job, are a paradox. Their rituals reflect the degree to which patriarchal gender relations that promote the image of the male breadwinner are not only interwoven into the fabric of the brotherhood, but are consciously reproduced as a blueprint for maintaining a sexual hierarchy at work and at home.

Construction unions in New York City, particularly Local 3, still maintain a great deal of influence over hiring practices in the industry (50 percent of New York City electrical workers are unionized). The unionized electrical contractors depend on Local 3 to supply them with highly trained journeymen and apprentices. The JIB runs a hiring hall where workers receive "tickets" for job assignments. Despite the power the union has over the labor supply, it is the contractors who ultimately determine hiring and employment practices in the industry. My informants generally agreed that if the contractors were serious about complying with affirmative action goals for women and minorities, the union would comply as well. Unfortunately, neither construction contractors nor building trades unions from across the various trades see the integration of women as beneficial into electrical work or the industry as a whole.

A collusive relationship like this is a significant variable in determining the speed at which a unionized craft occupation integrates women. As in other industries, it is management that holds the key and can unlock the door of bigotry in the workplace. Knowing the employers in this study to be some of the most powerful and wealthy contractors in the country, I was curious as to why employers were so disturbed about the entrance of a handful of women into the field. And why does Local 3, which has a relatively good track record of recruiting ethnically diverse tradesmen into the trade and union, fear a small group of tradeswomen in the union and on work sites? These were two simple questions which went begging in search of a method for an answer.

Anthropologists can no longer portray non-Western peoples with unchallenged authority. The ethnography of complex societies and institutions employ methods used in other social science disciplines to explain otherness and difference within the cultures of the West.[25] The ethnographer is compelled to climb down from the mountaintop of generalized theories to experience social relations on the ground. Thus this book is an amalgam of historical archival research, ethnographic methods of participant observation, and qualitative and quantitative analysis. The text ties the two dimensions of exclusion (that is, the organizational and the subtle) in order to provide explanations for why male and female workers are accepted or excluded at the construction job site.

I view the workplace and organizational culture of the electrical trade and brotherhood as contested, temporal, and emergent.[26] The industry and occupation are constantly changing and electrician brothers are trying to adjust. I never set out to do only a case study; in fact, electrical contractors are not the only building tradesmen to resist compliance with affirmative action. New work groups in the skilled trades such as women and minorities are more fiercely resisted in the higher-skilled mechanical trades such as carpenters, electricians, plumbers, and steamfitters, as compared to less-skilled "trowel" trades like laborers. However, it is in the unionized sector of construction wherein the greatest opportunities exist for decent work. Moreover, the unionized sector of the building trades is subject to greater governmental oversight than nonunion due to the necessary certifications required to operate union apprenticeship programs. If greater enforcement is applied from the federal and state level to monitor unionized contractors' compliance with affirmative action regulations, then there is cautious optimism for inclusion.

In general, tradesmen and their labor organizations have fought to exert control over the available labor supply to contractors and to achieve a "family wage." Since the late 1970s, women have been struggling to integrate with little progress. But it is in a mechanical building trade, like electrical construction, that it is most difficult. Furthermore, as previously stated, there exists an ample amount of differences between and among these highly skilled mechanical trades and their craft union brotherhoods to warrant not only an industry-specific but also a trade-specific study.

What are these differences? Local 3 is distinct because of (1) its highly evolved forms of fraternalism and brotherly rituals; (2) its tenacious and successful maintenance of formal craft apprenticeship traditions; and (3) its powerful influence over other locals in the IBEW, the IBEW itself, as well as New York unions as a whole. My goal with this case study is to shed light on broader issues of race, gender, and class-driven power dynamics that account more broadly for sex segregation in the labor market.

In the early 1960s, it was obvious that exclusionary hiring practices were emblematic of broader societal issues of poverty and inequality. Race riots and demonstrations on inner-city construction projects were commonplace, such as Harlem Hospital, the Downstate Medical Center in Bedford-Stuyvesant, and Woodhull Hospital in Brooklyn, but racial strife on construction sites and ethnic succession in metropolitan areas such as New York City are challenges that the electrical brotherhood has weathered. There are, however, new challenges building trades unions confront today: the advancing nonunion movement, new technological innovations coupled with management's push to streamline production in work crews, and today's globalization trends, all of which threaten to short-circuit traditional forms of union solidarity and brotherhood.

Another threat to the electrical trade unions are companies such as AT&T and the telecommunications industry who tempt skilled electricians who have been trained in rigorous union apprenticeship programs away from the unions to work as managers. In addition, the electrical brotherhood has recently faced antitrust lawsuits brought against it by the telecommunications industry that challenge the time-honored practice of a closed shop (union members only) on construction sites.[27] On the supply side, historically, women have been viewed as incapable and uninterested in building trades work. Paradoxically, black men and other immigrant and minority male workers are tolerated at a higher level than women across the board in the electrical industry and in other building trades as well.

In the face of employer brutality and deadly working conditions, the white working-class men of the building trades formed powerful unions to look after their interests. Whether or not they will be able to sustain this power will depend on their ability to adapt. In the electrical trade, the triangular relationship that fosters cohesion and collusion among male workers, the union brotherhood, and the contractors association unites mostly white men in a cross-class alliance to keep women and, to a lesser extent, minority men out of their ranks. For the union brotherhood as well as other building trades unions to survive, these patriarchal and exclusionary practices need to change.

The Participants

As an ethnographer, I set out not merely to invite respondents for interviews but to learn from electricians and their organizations. By gaining access to construction sites, electricians' homes, union halls, apprenticeship and industry meetings, as well as cultural, educational, and recreational events, I was able to survey the organizational structure of the brotherhood and contractors' association. In addition to this form of participant observation, I conducted in-depth interviews with male and female journeymen and apprentices, male union leaders, and electrical contractors.

More than four hundred people were interviewed for this book; three hundred were female electricians (journeywomen and apprentices) who entered the field from the 1970s to 2008. Of this latter group, 231 were of European American extraction, fifty-six were native black women, ten were Hispanic women, and three were Asian American. I also interviewed ninety-five male journeymen, of whom twenty-five were black (two of whom were part of the first class of African Americans to enter the Local), ten were Hispanic, eight were Asian American, and fifty-two were of European American extraction. Among the union leaders and activists, five top officers of the Local and JIB (all white men except for one African American apprentice director) were interviewed, including the son and grandson of the founder, the former education director at the Local, business managers, and ten contractors (eight males and two females) who supply most of the unionized share of electrical contracting for skyscraper work in New York.

I also interviewed former and current fraternal club presidents in the brotherhood, and when access allowed, the founders of specific clubs such as the Jewish Electrical Welfare Club, the Asian American Cultural Society, the Amber Light Society, Women Electricians, the Women's Active Association, and various counselors and officers of Camp Integrity (the Local's former summer camp for members' children) at Bayberry Land on Long Island. In these interviews, I had several aims. First, I wanted to include a substantial sample of women who had entered the trade from the first group to the last wave of entrants (i.e., from 1978 to 2008). Second, I wanted to obtain the views of the leadership of this craft union, its employers, and its rank-and-file workers regarding their thoughts about ethnic, racial, and gender integration. Even though my intention was to tell the story of women's heroic attempts to integrate the industry, I quickly became aware of the limitations of primarily using first-person testimonies and the importance of going further to see this integration from the eyes of male workers, unionists, and employers, in comparison with and contrast to the prior integration of other groups, such as minority and immigrant men. My sample included women of color, especially Latinas, African Americans, and Asians, in proportion to their numbers in the trade. Although interview questions were structured, I allowed a significant amount of time for respondents to veer off on tangents.

Restructuring my original goals allowed me to obtain valuable insights. Most interviews were tape-recorded and transcribed verbatim so that each interviewee's story could be read "as a whole" and also compared with others for differences and similarities of informants' opinions to the extent they might be shaped by such variables as race, age, class status, sexual preference, and education. In this way, I was able to sort out and analyze some of the most salient themes in both women's and men's experiences concerning sex and racial integration in this occupation, and to observe both the logic and contradictions in the primary research material.

Although the male journeymen were rather reluctant to speak with a researcher about their industry and union, the apprentices—both male and female—were more accommodating, seeing it as an opportunity to tell their side of the story as an underclass of workers.

Stepping inside the circle of power within the union brotherhood was quite another matter. The brotherhood operates like a prototypical military organization, with a strict chain of command. If I violated protocol, I would be shut out and my research shut down. The ability to conduct extensive interviews with union officials, workers, electrical contractors, and other relevant respondents was facilitated by the fact that I was a former professor of labor studies at the State University of New York (SUNY), where the Local to this day maintains its electrical apprenticeship program. This entrée allowed me to delve into some of the oral history presented in this work from the eyes of unionists, contractors, and workers who were willing to speak to me both on the record and anonymously, as appropriate. At the time I taught in the electrical apprenticeship college program at the Charles Evans Hughes High School in Manhattan, I was also able to conduct in-depth interviews that form the essential basis of the book with pioneers of the Local, especially with the late Harry Van Arsdale Jr.

These interviews with male and female electricians provided a comprehensive understanding of the problems and issues that each generation of female electricians encountered in this male-dominated industry and union from 1978, when the first wave of female electricians entered, until the beginning of the twenty-first century. Since the industry is predominantly male, I interviewed an equal number of men in the field, as well as union leaders and industry representatives, in order to construct an understanding of their collusive relationship regarding women's attempts at integration.

In addition, it was necessary to compare the entrance of women in the late 1970s with the entrance of other vulnerable male workers, such as black and Hispanic men, into the IBEW in earlier periods. In order to achieve the aims of the book, a complex set of actions and constraints were considered: (1) the history of the brotherhood and its organizational structure prior to women's arrival on the scene; (2) the historical formation and transformation of male fraternalism in the electricians' trade and its relationship to a culture of exclusion; (3) the nature of the labor-management relationship craft unions have had with employers' associations and the ways in which this has shaped men's and women's experiences in the trades; (4) the comparison of women's attempts at integration with those of earlier groups such as black and minority men; and (5) the interrelationships of race, gender ideology, and sexuality as patterns of unity or division among workers.

The choice of relevant scholarship for any book will no doubt always bring criticism about what was left out. This book is no exception. The vast literature on the social, psychological, and supply-side aspects of women's experiences in nontraditional jobs and apprenticeship, as well as the rich and colorful oral histories by authors such as Molly Martin, Susan Eisenberg, and Jane Schroedel, are indeed important for this work. Nonetheless, the central focus of this study of electricians differs somewhat from these prior works.[28] Instead, prior scholarship in historical sociology, labor history, political economy, and anthropology are more influential.[29] In this book I attempt to demystify the process of labor as a fully human process and, in place of overly deterministic views of gender and race segregation in the workplace, workers are portrayed as actors who continuously shape their own identities within specific contexts of privilege and opportunities as well as by the larger social forces in society.

Historical works on the history of the building trades were drawn from W. J. Rorabaugh's account of the devolution of craft apprenticeship, its race-driven aspects, and links to the patriarchal family in America.[30] Sean Wilentz's account of the degradation of craft labor in New York provides an important historical context for this book. Wilentz portrays the degradation of craft trades before and after electricians entered the labor scene in late nineteenth-century New York City.[31] In addition, David Montgomery's work linking manliness with workers' and labor organization resistance to unfair labor practices by employers in construction helped to shape my understanding of craft pride and the role gender solidarity plays among today's electricians.[32] For an international comparative view, Eric Hobsbawm's work on pre-industrial journeyman and apprentice craft culture provide a reference for some of the attitudes found among male electricians today.[33]

Historical case studies of male and female workplace culture were invaluable such as the traveling tradition of unionized cigar makers by Patricia A. Cooper,[34] Ileen A. DeVault's study of gender relations in craft unions and strikes,[35] Alice Kessler-Harris's work on the social meaning of women's devalued wages at work, and its relationship to their subordinate status at home.[36] In addition, Kessler-Harris's work on race and gender integration and public policy influenced me to look comparatively at processes of integration as simultaneously distinct from but related to one another.[37] As one black journeywoman electrician stated, "race, like torque, compounds the problems that women of color encounter." More recently, Nancy MacLean's excellent account of race integration in general into the building trades is relevant to the inclusion of men of color into the brotherhood's fraternal clubs, despite the strong degree of racism in society.[38]

Sociologists such as Ruth Milkman influenced my understanding of employers' customs regarding hiring patterns in auto and electrical work. British researcher Cynthia Cockburn enriched my understanding of rapidly changing technology in craft trades and the role it plays in eradicating or reinforcing separate work spheres between women and men.[39] But it is Mary Ann Clawson's seminal work on the transformative role of fraternalism in shaping gender relations at work and at home in eighteenth-, nineteenth-, and twentieth-century America that is the main literary inspiration for *Live Wire*.[40]

According to Clawson, social fraternalism in the nineteenth and early twentieth century challenged the concept of women's superior moral values by creating and preserving male authority outside the nuclear family in a "men's world of virtue." Free-labor ideology, another aspect of fraternalism, addresses non-economic relations; that is, men rising above the cruelties of business to assert solidarity across classes. Thus employers and workers join together in fraternal bonds. Gender, not class, is the solidarity at the root of social fraternalism and the nexus of social relations at work and in leisure. There has never been a more compelling case of the transformational importance of fraternalism and its role in sex segregation than the modern-day electrical brotherhood.

The electrical brotherhood, or any union for that matter, is not directly descended from early modern European fraternities or nineteenth- or early twentieth-century fraternal trade organizations. But it has appropriated, reinterpreted, and selectively transformed the fraternal notions of those organizations.[41] By wiring together formal and informal cultural forms of male bonding and gender solidarity for purposes of organizational efficiency and commercial expansion, the electrical brotherhood and its joint industry board provide the nexus that privileges white male workers, unionists, and employers to the exclusion of women.

Milkman, Cockburn, and Clawson, among others, have in different ways illustrated that management's gendered patterns of recruiting, hiring, and promoting may also influence a sexual hierarchy in workers' collectives like union brotherhoods. What these authors have left unexamined are the ways in which a collusive relationship among male workers, unionists, and employers once used as an effective means to combat employer exploitation, operate in the twenty-first century to the exclusion of women.

Live Wire is my response to this. Chapter 1 surveys the background history of the electrical construction industry, the union concept of "brotherhood," and women's presence in the trade. Chapter 2 takes a close look at Local 3, its ethnographic history, organizational structure, and fraternal character forged by resistance and accommodation to the free market's laws of supply and demand. Chapter 3 details how women secured a foothold in electrical construction and the brotherhood in the first place, and how they have struggled with the union to maintain it. Chapter 4 takes the reader through a typical electrical construction workday, showing how the dynamics of gender politics color the interactions of construction workers on the job and at home. Chapter 5 focuses on the powerful factor of race on women in the industry. Chapter 6 lays out the dramatic story of the ongoing attempt to institutionalize women electricians' interests through a succession of women's sororal clubs—and the complicated question of whether these clubs should work as auxiliaries or adversaries to the male union leadership. In the concluding chapter, the experiences of women electricians are recapitulated and analyzed for broader insights into the role that fraternal customs; workplace traditions; conceptualizations of the family; the intersectionality of race, gender, and ethnicity; and occupational culture play in shaping patterns of workplace inclusion.

The concluding chapter also suggests ways in which present-day public policy can influence greater equality of opportunity in the industry and union brotherhoods to get women down to the job sites.

1
Brotherhood: The History

O ff a rambling road on the tip of Long Island, for more than eighty-five years a sprawling mansion stood over a white beach facing the Atlantic Ocean. Designed in the style of an English manor house, completed in 1919, and dubbed Bayberry Land, the 300-acre estate served as the summer home of Guaranty Trust Company president Charles H. Sabin and his bride Pauline Morton Smith, heiress to the Morton Salt Company ("When It Rains, It Pours").[1] Sabin, described in an obituary in the *New York Times* described him as "one of the most prominent figures in American finance,"[2] died in 1933. Pauline died in 1945. A developer bought the estate in 2001, demolished the mansion and outbuildings, and in early 2006 opened the Sebonack Golf Club, styled as the most expensive golfing facility in America. But for almost fifty years in between, this Southampton estate was owned by Local 3 of the International Brotherhood of Electrical Workers (IBEW).

Influenced by the vision of Harry Van Arsdale Jr., the longtime business manager of Local 3, the union bought the property in 1949 and restored it. A union convalescent facility opened in 1952. Beginning in 1957 the union ran educational programs at Bayberry Land, and Camp Integrity, a summer camp for children of union members, was established in 1971.[3] It was an act of foresight, strength—and defiance. Against the received wisdom of the day, Van Arsdale held that union members and their families should enjoy the finer things in life. As he said, "There should be access to those things that are not just for workers."[4]

July 1999

It is 5:00 PM, the close of a long training session on preventing sexual harassment.[5] Fifty male electricians, business agents, shop stewards, foremen, and union officials have been sitting in my class in a large auditorium in the Bayberry complex for over five hours. We have been reviewing the contract and compliance obligations of both contractors and unionists to prevent sexual harassment on the job. The discussion has been contentious at times. Now that class is over, the men say it is time to put aside the "the battle of the sexes" and break out the beers. On the beach, my students and I set up the frying equipment and begin the ritual that is one of the brothers' all-time favorites. While the waves break and roll, we fry a turkey for dinner. I feel a strong sense of solidarity with these men.[6] The scene that night reminds me of my father's annual Operating Engineers' picnic, which I attended as a child. The electricians are speaking a language I know; it is like coming home.

The brothers on the beach find kinship and comfort from their strong relationships with one another. They bond together to strengthen their precarious status as construction workers in a free-market economy. They are well aware that Local 3 separates them from the much poorer (often immigrant) construction workers who are not unionized. They know that the union gives them access to the American dream and a relatively stable middle-class existence. The Brotherhood is a vehicle for economic security from cradle to grave, so they are loyal to the union.

Male gender solidarity is one of the union's customs and traditions. It played a role in the union's struggle for survival in the early twentieth century. Today, Local 3 remains an institution in which gender solidarity prevails. Some of these very men I feel I know so well are quite capable of inexcusable behavior toward women. They may revel in their camaraderie with one another, but some may also despise and harass women in the industry. They sit idly by while contractors and unionists trample on women's rights and dignity. Some even endanger the lives of female workers by denying them training and then assigning them dangerous jobs. The Brotherhood makes it difficult for women to join its ranks.

How did this situation evolve? Why is it that when many fields have achieved gender integration, the construction trades have not? To answer these questions, we must consider the history of organized labor in general, and craft electrical work in particular.

The Market Economy

The transition to a market economy in sixteenth-century Europe brought a major change in the labor system: men now worked outside the home using materials and machinery that they did not own, and their value was determined by the price they could command for their labor. By contrast, women's work in the home was not generally allocated a market value. This disparity in the monetary value of work gave male workers a sense of social superiority as money became the primary, then the only, measure of worth.[7]

Craft workers, the majority of whom were male, organized themselves into guilds and secret crafts societies, the precursors of craft unions. Although women were denied anything like full membership in such workers' fraternities as the Knights of Pythias and cross-class social organizations such as the Masons, they were sometimes organized into subordinate auxiliary organizations.[8]

By the end of the eighteenth century, men's gender rights were based more on contract than on custom. In the new civil society based on a market economy, men related to other men on terms of fraternity, while relegating women to subordinate positions in the household. Fraternal societies composed of white men controlled access to commercial networks and craft occupations, creating, as Irish feminist and writer, Mary Wollstonecraft, proclaimed, "a brotherhood that excluded sisters."[9]

From Fraternal Orders to Trade Unions

"Their idea seemed simple enough at the time: If building trades unions intended to improve conditions, prevent conflict, and retain control over their work, they would have to join forces in a national organization."[10] Being white and male were the unifying factors among the union founders, who envisioned a plan to unite the "basic" trades, annihilate rivals, and regulate jurisdiction to produce "harmony, equity, and rightful ownership in the building industry, . . . [believing that] the internationals would stand as a unit one with the other. Here an injury to one would be . . . the concern of all."[11] The fraternal foundation of an industry that is highly contingent, geographically dispersed, culturally diverse, and susceptible to violent sabotage is part of the intriguing story behind the building of skyscrapers. The masculine underpinnings of the industry help frame our understanding of why the union brotherhoods are organized as they are and why change regarding the entrance of women is so slow. "Unless we hang together, we will hang separately," warned William J. Spencer, a labor leader who was secretary-treasurer of the Building Trades Department of the American Federation of Labor (AFL).[12] However, AFL leaders actually meant that "an injury to one would be . . . the concern of all [white men]."[13]

Fraternity is the very foundation upon which proto–trade unionism was built.[14] As stated previously, early modern American craft unions evolved from racially exclusive, strictly fraternal organizations to social organizations that joined journeymen, master craftsmen, and workers across the trades (such as the Knights of Pythias, the Masons, and the Fraternal Order of Workmen). These fraternal societies included apprentices, journeymen craftsmen, and master mechanics or owners. They served as venues for job networking, negotiating immigration problems, and developing credit unions. They also served valuable familial and community functions in an unstable economic environment.[15] In general, mutual voluntary associations provided a safety net to male workers who were exposed to the brutal vicissitudes of an uncertain market. For example, the Ancient Order of United Workmen, the fifth-largest fraternal order in the country, with 357,000 members, was the first fraternal order to offer life insurance, in the late 1890s.[16]

Trades-centered labor organizations such as the Grand Forge of the United States; Sons of Vulcan, an early ironworkers' union; and the Knights of St. Crispin, the large and militant shoemakers union, were also modeled on the Masons.[17] Some organizations, such as the International Association of Mechanical Engineers, abandoned the outward trappings of fraternal orders and simply called themselves "unions"—while hewing to codes of secrecy and rituals that would make the Masons proud.[18] For laborers seeking leverage against both employers and nonunion arrivistes, these private rituals helped create to solidarity and to unite workers in bonds of loyalty and trust.[19]

In the political and economic climate of rapid industrialization and scant business regulation after the Civil War, brotherhoods flourished. Union locals were typically called lodges, and it was standard practice to call fellow unionists "brothers" and to sign letters "Fraternally yours." The idea of the union as a fraternity remains the dominant model in electrical craft work and construction brotherhoods to this day. Early electrical craftsmen organized themselves into grassroots fraternal organizations that served many functions and played an important part in their members' lives:[20] they were at once scientific and cultural societies, commercial networks, benevolent societies, and recreational clubs.

These organizations were sometimes welcomed by employers, who appreciated having an institutionalized system for resolving differences between themselves and their employees. As I have said, the construction industry was (and is) highly contingent, geographically dispersed, culturally diverse and susceptible to violent sabotage. Building skyscrapers and building the construction brotherhoods went hand in hand, and masculinity was embedded in both. Fraternal organizations helped circumvent strikes and lockouts and helped the industry avoid general chaos.[21]

For these fraternal organizations in the building trades to improve conditions, prevent conflict, and retain control over their work, they had to join forces in a national organization.[22] And because the various brotherhoods were overwhelmingly male, gender solidarity was the uniting factor that would produce "harmony, equity, and equality."[23] Understanding this culture helps us comprehend why it remains so hard for women to enter the ranks of union electricians.

Electrical Workers

Throughout the twentieth century, the construction industry and the power-generation industry grew together to become twin titans of the American economy. Advances in building technology such as reinforced concrete, steel framing, and elevator systems helped make the skyscraper the symbol of American industrial might. Concurrently, commercially produced alternating current made possible the widespread use of electricity, thus modernizing the industrial power industry. The symbiotic rise of the construction and power industries made skilled electrical labor a valuable commodity.[24]

Construction electricians adopted a craft-union model of uniting members along the lines of skilled trades work, such as ironwork, carpentry, or masonry, or specialties like plumbing or painting, and centered on an apprentice system and father-to-son sponsorship. In the tradition of the benevolent societies and their ideology of manhood, the electricians' union fought to secure a "family wage" to enable men to provide for their wives and children. The union also provided benefits to help workers through times of recession and layoff. In addition, it offered cultural activities to develop solidarity among members. The apprentice system also gave workers a valued link to the past. The ideology of labor brotherhood guaranteed to worker and boss alike a very desirable end: stability.

The ever-expanding reach of the market into the economic structure of society transformed or obliterated many traditional customs and practices, but not all. The practice of father-to-son unionism bestowed on late nineteenth-century workingmen a measure of security: their sons had a birthright to training in a skill. That right remains an important component of working-class dignity, pride, and manliness. Van Arsdale put it this way:

> A man was some type of craftsman like a carpenter or a bricklayer, and he had achieved this skill. He had gone through an apprenticeship—he had worked a lifetime—he had no reserves. What was his estate? What did he have that he could give to his children? One of the things he had, and sometimes the only thing, was his skill. So it was practical that his sons should come into the union as apprentices ahead of anybody else. I might be teaching your son and you might be teaching someone else's son—we are in fact collectively imparting to our children what we have achieved in our lives.[25]

In the face of unemployment, recession, and depression, early trade union brotherhoods viewed kinship as the organizational means through which craft pride could be preserved and transmitted. For their part, employers eventually came to recognize the benefit of the union brotherhood to provide them with a flexible supply of workers in an industry in which labor demand was notoriously erratic.

The Brotherhood of Electrical Workers

The first labor organization in the electrical trade was the National (later International) Brotherhood of Electrical Workers, founded in St. Louis, Missouri, on July 19, 1890.[26] The IBEW did little to improve either wages or hazardous working conditions in both the construction and power industries, and the depression of 1893–1897 almost destroyed the young Brotherhood completely.[27] The early twentieth century was chaotic, with unsanctioned strikes and a rebellious rank and file that frequently disregarded both contracts and union leadership. Membership turnover was extraordinarily high. In 1903, for instance, ten thousand of the union's thirteen thousand members were new recruits.[28]

Relief in the form of stabilized membership rolls and dues came from an unexpected quarter, one that the union brothers had a very hard time acknowledging: women. Skilled work in construction made up only half the jurisdiction of the IBEW. The other half was factory-type work in the electrical industry, including power generation, the rapidly developing telegraph and telephone business, and manufacturing.[29] Here the skills that gave construction workers the power to enforce their select brotherhood and maintain its exclusive rituals were irrelevant. The jobs were viewed as easy, and employers turned to women as a cheap way to fill them.[30]

The IBEW had initiated a few women members in 1892, but their presence caused such dissent that the following year it restricted membership to men.[31] By 1895, however, the increasing employment of women in these proliferating low-wage jobs forced the union to readmit women, but at reduced fees commensurate with their paltry salaries.[32] Women helped buttress the union's organizational and financial strength. For a union with "Brotherhood" in its very name, admitting women was painful. As a letter in the IBEW's newspaper lamented, "If these fair creatures are admitted, they will have to hold office, then the brick will fall—Oh God! Deliver us from being ruled by a woman."[33]

Nonetheless, the women were strong unionists. During a 1900 strike of telephone workers, the female operators told employers, "We would rather starve than desert our brothers in their struggle for what they are entitled to."[34] The "hello girls"—telephone operators—were also organizing and striking successfully,[35] although it did not help their status within the union. Whenever a strike or organizing drive among women workers failed, men blamed the lack of success on the women rather than the employers—even if the general rate of failure was exceptionally low.[36]

In 1902, the IBEW's executive board announced that its official policy would be to organize women into separate locals.[37] Although the female-dominated manufacturing and the male-dominated construction divisions were different enough to warrant the development of separate jurisdictions by the union, historical evidence reveals that men consciously applauded having women in separate divisions.[38] As an added benefit, keeping the genders separate reinforced the male workers' demands for a family wage. Accordingly, they also proclaimed their authority and privilege over dependent wives, sisters, and daughters at home. Furthermore, if women worked, employers might be apt to argue that male wages sufficient to support a family were unnecessary. Even young unmarried male apprentices in anticipation of claiming their "estate" as journeymen, define manliness as becoming the male breadwinner in the family.

Allowing women to enter "male-typed" work jurisdictions or to integrate male-only divisions would undermine the very source of craft-union power: the ideology of brotherhood itself. Their very sense of themselves as providers would be under threat. Furthermore, any foray of women into their jurisdictions would threaten the pecking order and the political power within the brotherhood.[39]

In addition, notions of brotherhood not only went far beyond the concern for the individual workers and their families but to the communities where workers lived. Cradle-to-grave benefits established by the brotherhood such as medical care, covering burial costs, and providing retirement funds by craft union custom and tradition linked work to home to the community at large, thereby supplanting earlier American mutual voluntary fraternal societies.

A Short History of Local 3

Local 3 of the IBEW is the umbrella organization for most electrical workers in the greater New York metropolitan area and is the subject of the ethnography that follows. Local 3 is unusual because it includes both a craft and an industrial section—the first traditionally reserved for men, the second for women. At the present time, although membership in Local 3 has waned somewhat since the economic recession of the early 1970s, its current membership of more than 35,000 makes it the largest building trades local union in the United States, and about 100 times larger than the average local within the IBEW. By virtue of its size, and quite apart from the principle of local autonomy, Local 3 exerts enormous influence within the IBEW which is the largest international union of workers in the electrical industry worldwide (approximately 750,000).

Electrical craftsmen first organized the Electrical Mechanical Wiremen's Association of the Knights of Labor in 1887.[40] This was composed entirely of men of northern European descent (Scots, Germans, and Irish).[41] After an unsuccessful strike against Western Electric, Association members gathered in a Masonic Temple in Queens, New York, and voted in 1891 to affiliate with the AFL.[42] By 1893, the Local was a full-fledged member of the AFL (as well as the Knights of Labor) and was negotiating with the Edison Company.[43] The young Local, uncertain of the nature of its affiliation with the

AFL, sent delegates to the IBEW's first convention but refrained from official association until February 1900.

This early period in the electrical industry (and in construction as a whole) was characterized by weak contracts, years of strikes and lockouts, the absence of strike benefits, and internal strife.[44] "Due to the anarchy of the industry itself and the anti-union spirit of the time," observes historian Maurice F. Neufeld, the union "was forced to create for itself a sense of solidarity against the world, the employers, the forces of law and order, the judiciary, the Communists, its own former officers, its own international, and other unions."[45] The concept of brotherhood as a basis for class and ethnic solidarity was what kept the fledgling organization alive through this bitter period.

Keeping their own economic interests at the forefront, employers initially resisted workers' organizations. Later, however, they came to embrace unions as mechanisms to establish a modicum of cooperation with labor so as to avoid the losses that came with any strike.[46] The evolution of the modern day Joint Industry Board of the Electrical Industry (JIB) and the proliferation of management-labor committees to redress labor conflicts across the construction building trades speak to the desire of employers to avoid delays in the timely completion of projects, thereby holding down costs of loans and extra material costs for building work.

Early in the twentieth century, contractors and building owners systematically tried to break Local 3. The next generation remembered the struggle. In a 1985 interview, Harry Van Arsdale Jr., the son of Local 3's founder and himself the business manager for fifty years, related that for a full thirty-three months starting in 1903, owners refused to hire Local members at "shape ups"—the sporadic gatherings of day laborers or on-call workers where employers made their work selections for the day. Informers were paid to identify union members, who were then dismissed. As Van Arsdale recalled, the resultant hardship was so severe that "only 400 of the Local's 1,800 members refused to renounce their union membership."[47]

Bitter strikes continued to strain the Local's limited resources as labor-management conflicts raged throughout the 1920s.[48] But the Local survived. The sons of the 400 "pioneers" (as the Local calls them) eventually became the dissident rank-and-file group called the "Loyal 100." This group would radically transform the Local in the early 1930s, instituting sophisticated governance procedures, a paramilitary organizational style, guided democracy, and kinship networks.[49] They refashioned clubs such as the Masons into voluntary fraternal orders that could justify the integration of ethnic layering reflective of the immigrant workforce of the day. To this day, descendants of this group have remained in power, providing the fourth generation of leaders in Local 3.[50]

Van Arsdale also revealed that in the early twentieth century, construction electricians could not count on regular work.[51] The seasonal nature of the building industry as well as contractors' anti-union attitudes made employment insecure. To hide the fact that they were scabs, Western Electric workers crossed picket lines with briefcases "from which they extracted their soldering irons," stated Van Arsdale.[52]

Local 3 gained absolute jurisdiction over manual work in the construction of movie houses, and its members blocked nonunion engineers from working, insisting that union electricians do the job.[53] During a period of rapid technological progress, the Local shrewdly devised strategies for diversifying craft work. While other electrician locals lost out to the engineers over the installation of sound equipment, Local 3 protected its jurisdiction by maintaining a high standard of training and education. As a result, the union electrician was often the best man for the job.

In the 1920s, Local 3 took an active role in New York City politics and in the larger labor movement. It also did something unique among building trades' craft unions. Though originally a craft local, it expanded its reach to lower-skilled electrical manufacturing workers as well.[54] Construction craft unions generally had little interest in the rapidly growing industrial workforce, whether male or female.[55] According to Van Arsdale, however, by the late 1920s Local 3 was organizing both horizontally, reaching out to electrical tradesmen, and vertically, recruiting elements of the industrial workforce even if they were only distantly related to electrical construction work.[56] Inspired by Lincoln Steffens's *Autobiography* (1936), the journalist and son of a wealthy California businessman who later in his career developed radical political views,[57] the Local set out to organize switchboard operators and people who made lampshades, Bakelite switches, and cables.[58] In the wake of the federal Wagner Act of 1935 (which required employers to negotiate with unions), the Local also set out to bring in even more workers from these expanding industries. Business agents, mostly men of northern European descent, organized women from various immigrant backgrounds—first Irish, Polish, and Italian, and (after the Depression) Puerto Rican, African American, West Indian, and Haitian—even as the AFL leadership assumed women could not be organized. But Local 3 did the job with no trouble. Women workers at lamp and shade shops were viewed as different, and difficult, mostly because they were poorer, Van Arsdale recalled. "They (Local 3 male organizers) would go into the lamp and shade shops to recruit these women, and they would come to union meetings with piecework piled in bundles high on their heads like in a foreign country."[59] But he was convinced that these women would be undeniable assets to the union—"a great bunch of people."[60] Van Arsdale relished the irony of having a Congress of Industrial Organizations (CIO)-type industrial shop in an exclusionary AFL union.

With regard to women, Local 3 mirrored the structure of the IBEW. Just as the Brotherhood organized female switchboard operators into a separate unit, Local 3 created separate divisions for different kinds of work, often already segregated by sex by the employer.[61] Thus the union preserved male solidarity and the gendered hierarchy of the brotherhood while assimilating women as union members—as long as they did not stray from their lower-skilled job categories. By the mid-1950s, Local 3 had created a Women's Active Association (WAA) composed solely of manufacturing workers—that is, a sororal society to which women from the segregated manufacturing sector belonged. Among the tasks relegated to the WAA were overseeing elections in the Local and reporting problems with employers in the electrical manufacturing shops to the Local's business managers.[62]

This development took place in a period of increased stability for the Local, which was achieved by locking into place the consolidated fraternal structure under which the organization still thrives.[63] The old Loyal 100, reconstituted as the Committee of 100, was reserved for members carefully screened for capacity, loyalty, and dedication. It became what is called a "commando" group, working under the business manager to enforce union interests and to rid the Local of Communist influence and corrupt practices under the former regime of business manager Harold Broach.[64] For these men, union activism was an "estate" inherited from their fathers and uncles, along with their union jobs and union positions.

Labor prospered during the economic growth of the 1950s, and it was at this time that Local 3 acquired Bayberry Land for the use of union members and their families.[65] The Local created a veritable workers' community in which social clubs, often organized on gender, ethnic, or religious lines, gave members and their families a way to establish class solidarity.[66]

While European American tradesmen, especially the descendents of the Dutch, Irish, and German, maintained their hegemony over the Local and its powerful political machine, direct action at construction sites and clandestine meetings of Jewish electricians in New York City in the construction and film industries prompted the formation in 1923 of a Jewish club, the Electrical Welfare Club. The sanctioned club signaled the acceptance and subsequent integration of male Jewish electricians into the brotherhood's prestigious Division A.[67] Owing much to the battles fought and won by these Jewish electricians in the 1920s, minority men and women in other divisions of the union founded clubs as well.[68] Although the Local did not sanction minority clubs until the 1950s, as far back as the 1940s the leadership took a progressive stand on issues of racial and ethnic equality, but only among male workers.[69] The Latino Club, now known as the Santiago Iglesias Society (in honor of the revered Puerto Rican union organizer), was founded in 1948 after the Local organized a study trip to Puerto Rico.[70] Concerned about the large influx of unorganized Puerto Rican immigrants and other Caribbean women workers into the manufacturing industries in New York City, Local officials met in Puerto Rico, the U.S. Virgin Islands and Jamaica with host labor leaders. They explained how poor working conditions and low wages for manufacturing work on the islands led workers to look for jobs in the light manufacturing shops that Local 3 had organized in the 1930s and 1940s. Hispanic and black Caribbean women workers were replacing earlier European (e.g., Italian, Polish, and Irish) immigrants who were seeking better employment opportunities as office workers. These women of color developed an ethnic niche in light manufacturing shops and supply houses in the electrical industry, such as in the Levittown factory in the Greenpoint section of Brooklyn.[71]

The official story surrounding the 1958 creation of the African American club, the Lewis Howard Latimer Society,[72] traces its origin to a black journeyman whose identity remains unknown in the Local. This construction division member wished to honor Latimer, an African American who was Thomas Edison's assistant in his New Jersey laboratory and a pioneer in the electrical industry in his own right.[73] Conflicting views abound for the reasons the Local sanctioned the club. According to interviews with a focus group of black male electricians, the club was founded as a way for the union leadership to co-opt dissident black members and more fully integrate them into the Local. Interviews with this same group reveal that some regard the club as merely a public relations arm of the Local. Whatever the real reason for the Lewis Howard Latimer Society, it is certain that the club has always lacked the formal and underground links to the political power and commercial information in the city that make membership in the other European American clubs so desirable.[74]

Each club had a business agent who reported to top Local officials and who served as a power broker between club members and the central union and Joint Industry Board administration. In some cases, officers and members of powerful older clubs with ties to older Local 3 union regimes, such as the Cornerstone Club and the Masons, were simply replaced by "commando" members of the Committee of 100.[75] Thus the predominantly white European male Cornerstone Club split into two fraternal clubs dominated by Catholics—the Allied Club and the Catholic Council—perhaps the two most powerful fraternities in the union today.[76] The Masons maintained their club, the Electrical Square Club, but its power was significantly reduced during the Van Arsdale regime, which limited members' activities to social outings, dinners and dances, and philanthropic work.[77] Van Arsdale and his loyal "commandos" dismantled the Masons' control over the Local, claiming that the club contained elements of the earlier corrupt Local 3 regime of former business manager Harold Broach, Van Arsdale's predecessor and rival.[78]

In the 1970s, the union faced a new challenge: women wanted to become electrical construction workers. This would prove to be one of the most challenging confrontations in the history of the Local, rivaled only by the growth of nonunion work.[79] Two groups of women had this goal: highly educated females from women's colleges and the feminist movement, and low-income minority women. In response, male electricians and their union leaders redoubled their commitment to the ideology of "brotherhood," "equality," and "fraternity" as they had come to define it. With the Local's gender identity written into its very genetic code, the craft workers of Local 3 fiercely guarded their privileges. And Local 3's customary dual structure of separate jurisdictions and divisions of male skilled workers and female factory workers made it especially difficult for female electricians to enter the trade. Since the Local traditionally organized women from low-wage work into less powerful segregated divisions, women electricians entering the elite construction division would cause the "Brotherhood" to be perceived as a misnomer and put in jeopardy the male electricians' expectations of a family wage at work and "king of the castle" status at home. A journeyman will have many employers and contractors over the course of a career as an electrician. Unlike employees in most industries, construction electricians, and workers generally, are temporarily employed until the building project is completed. They then move on to different job sites throughout the city, region, and at times, the country and abroad. This work relationship sharply contrasts with the traditional one employer/one employee relationship at a permanent work site characteristic of most industries and occupations. Consequently, building tradesmen developed fierce loyalty to their unions and a strong camaraderie built on male gender identity akin to the bonds of solidarity men develop in the military. The electrical brotherhood forged these bonds of loyalty and solidarity in the late nineteenth to early twentieth century as a formidable challenge to employers' brutalizing treatment of workers. But Local 3's polarized views of the roles of women and men at work and home were sharply reinforced in the 1950s when the brotherhood reached its ascendancy and power. Accordingly, even a handful of women entering the construction division in the late 1970s threatened "men's places" in the union and the institutional arrangements of male networks equated with the brotherhood's ascendancy to power.[80]

Today, Local 3 remains a segregated organization. More than thirty autonomous divisions represent various trades, from construction workers to cable splicers, from lampshade manufacturing workers to switchboard operators. More recent divisions include secretaries, clerical workers and maintenance workers. The elite and powerful construction division, itself composed of different classes of workers, is the most influential division in Local 3. Within it, the construction electricians of Division A are the elite of the elite. More than any other union members, they are the ultimate guardians of the traditional ideology. They oversee the apprenticeship system, which continues the tradition of fraternalism. They are responsible for developing union solidarity among new journeymen. Above all, the men of Division A pass down the traditions—the ideology, if you will—of the union, in which journeymen become craftsmen, who in turn train a new generation of journeymen—ideally, younger male relatives.[81]

Local 3's female manufacturing workers are the structural opposite of the elite Division A tradesman. From the start, despite the factors which may have influenced Van Arsdale to organize manufacturing workers, such as the influence of radicals like Lincoln Steffens or the Local's affiliation with the IBEW (which was mainly an industrial union), Local 3 male business agents organized women in the city's lamp and shade factories because they were a group in need of protection, and not because they regarded them as full-fledged union sisters. The leadership expanded on nineteenth-century definitions of brotherhood, claiming it was gender- and race-blind to the characteristics of these workers. The Levittown women's integration into the Local further buttressed earlier fraternal definitions of manliness by organizing these women workers in a paternalistic way. Despite a high degree of racism among white male electricians and society in general at the time, the tradesmen of Division A willingly accepted a higher dues assessment than these fledging women members and were assessed as a matter of "unselfish brotherhood." It was necessary to keep the electrical supply industry unionized, explained Van Arsdale to the members, so that Local 3 could more effectively thwart attempts by nonunion contractors to purchase more cheaply made electrical building products, which would in turn enable the nonunion sector to win contracts by outbidding unionized contractors.[82] In fact, propelled by Van Arsdale's visionary leadership, many former union leaders said in interviews that they believed the Local had to organize them as a last resort to protect related aspects of the electrical construction industry from going nonunion. As most vividly recounted by the Levittown women workers who subsequently became leaders in the manufacturing division of the union, the women welcomed the intervention: "We were desperate . . . conditions inside the shops at times became brutal especially in summers when the temperatures reached 100 degrees. The men from Local 3 presented us with an alternative to challenging the bosses alone. Women would faint on the line, and the ten-minute lunch break would find women in the powder room trying to revive each other while struggling to get a bite to eat before the factory whistle blew again. . . . All that changed with the strike and the union."[83]

Current Local 3 leaders deny any attempt either to reach out to women or to discriminate against them. They recall only gender-neutral attempts to organize manufacturing workers who just happened to be women. "No special issues ever came up," stated one union leader, comparing his experience organizing Levittown women workers to female electricians' current demands for benefits such as light duty during pregnancy and child care accommodations on work sites.[84] The union denies that it ever paid much attention to such issues, or that it makes particular allowances for the special circumstances of a woman's life. This is especially true for women electricians; the union believes that raising these issues on work sites will be costly to the contractor and a deterrent to the demand for union workers.

Interestingly, women members predominate in Local 3. However, following a pattern of sex segregation in the workforce, the leadership has persistently relegated women to second-class status, organizing them into separate and subordinate divisions according to its different jurisdictions. Even now, in the new millennium, the Local's ideology remains patriarchal and paternalistic. The way in which that ideology militates against inclusion of women is the subject of this book.

2
A Closer Look at Local 3

You go through some of the old neighborhoods, you know, you would see some of the signs on some structures where it was actually this ethnic group or that ethnic group. For the working man that was like a rich man's sporting club, or country or athletic club or whatever they call it—a social club.[1]

Today, Local 3 stubbornly clings to the policies, practices, and administrative structures that helped it succeed in the twentieth century. With over thirty-five thousand members, including thirteen thousand in the elite construction Division A, Local 3 has developed a large, hierarchical governing structure. A full-time business manager, elected every four years, runs the day-to-day operations, but the Committee of 100, whose membership is handed down from father to son, is a permanent feature of the brotherhood and guarantees its members organizational continuity.[2] As mentioned in Chapter 1, the Committee was formed from the Loyal 100, the union activists who supported the policies pushed by Harry Van Arsdale Jr., then in his thirties, son of an electrician, a member of Local 3, and just beginning his long career as Local 3's business manager.[3] In 1933, having succeeded in toppling Harold Broach, the Committee of 100 became the top tier of the organization, directly under the famed Business Manager Van Arsdale.[4] As stated earlier, Neufeld observed that, "every candidate for membership on this 'commando group' is carefully screened for capacity, loyalty and willingness to devote countless extra hours to the job of saving the union. It adopted for its own use the American Communist Party strategy of holding caucuses and planning programs in advance of union meetings, and each member was instrumental in the art of self-defense."[5]

In an era of strikes, police brutality, corruption, industry anarchy, and hostile employers, Local 3 used multiple strategies to build a strong organization. Most important was union discipline. This was realized through internal mechanisms that guaranteed a rotation of elites among the Van Arsdale regime; those mechanisms are still extant. For example, the fledging Local leadership built strong ties to politicians and to contractors who were former Local 3 electricians and created divisions along the lines of work jurisdictions, as well as fostered new fraternal social clubs to overcome the challenges posed by ethnic layering in New York City and the growth of nonunion work.

Unlike other craft unions of the day, Local 3 established strong ties with municipal organizations and local politicians.[6] The key to stable labor relations in the electrical contracting industry, however, was the negotiation in 1939 of a "Voluntary Code of Competition for the Electrical Contracting Industry of New York City," which led to joint union-employer management of a pension fund, and in 1943 to the establishment of the Joint Industry Board of the Electrical Industry (JIB).[7] Composed of Local 3 leaders and owners of the major electrical contracting firms in New York City (most of them former Local 3 journeymen),[8] the JIB ultimately regulated hiring practices, administered pension and benefit funds, and ran educational facilities and Camp Integrity summer programs for members and their families.[9] Although the practice of establishing joint industry boards is not unique to Local 3 (or strictly the construction industry), the extent to which the JIB successfully managed and administered a broad potpourri of benefits and services to its diverse and geographically dispersed union members, and its success in compelling New York City electrical contractors use union labor, set it apart from other joint union-employer efforts.[10]

Local 3's internal organization is based on work divisions defined by occupation or industry—and also by social clubs based on gender, ethnicity, religious solidarity, and geographic location.[11] But the story of the social clubs in Local 3 did not begin with the Committee and Van Arsdale's takeover of the union. The social clubs actually predated that leadership but were limited to two main influential clubs formed along religious lines: the Central Colony Club (predominantly Catholic) and the secret society of Masons Electrical Square Club (predominantly Protestant). Once in power, Van Arsdale and his "commandos" toppled the officers of these clubs, master electricians who were union leaders and loyal to the former regime. Instead of entirely dismantling the clubs, Van Arsdale limited their function to social gatherings and philanthropic activities, rendering them politically impotent.[12] He continued, however, to model the brotherhood along the lines of this prototype fraternal structure and established new clubs that served to extend and enforce the power of the Loyal 100 cadre. Van Arsdale viewed the clubs as useful administrative tools that allowed the union to organize more broadly and communicate more effectively with the growing cultural heterogeneity of workers in the city, a challenge to building union solidarity that is relevant to this day.[13] "If you want to see what a union should look like," remarked Van Arsdale, "just take the subway."[14]

In 1933, Van Arsdale and his Committee of 100 began consolidating his power through the club structure.[15] The leadership created two new top-tier fraternal clubs, the Allied Club and the Catholic Council (both of which are predominantly Dutch and Irish Catholic) and installed officers loyal to Van Arsdale. To this day, with their commitment to humanitarian activities, city politics, and loyalty to "God and Country," and with their ties to municipal, state, and national politicians, the two clubs remain the most powerful fraternities in the Local.[16]

Van Arsdale faced the daunting task from the 1930s through the 1950s of keeping electrical building trades work unionized. He knew that this meant increasing the number of Local members generally, as mentioned in Chapter 1, and using the political advantage a large membership offers to the advantage of tradesmen as well. Enlarging the Local to members outside of the elite Division A electricians was central to the brotherhood's ability to flex its muscle with influential Democratic and Republican politicians regarding the maintenance of electrical standards and building codes to thwart de-skilling of the trade. The visionary leader accomplished this by setting out to organize every aspect of electrical work needed for skyscraper building work. But Van Arsdale had to confront a paradox. While a growing union membership meant costly administrative oversight, the women workers Van Arsdale targeted as new members (from Levittown and other electrical supply factories) were all low-wage earners and could not afford to pay high union dues. Compounding this challenge, increasing prosperity after World War II allowed for greater geographic mobility on the part of electricians and their families, making it possible for them to move away from the neighborhoods that had bound them together as workers and neighbors.

In Van Arsdale's eyes, this threatened the solidarity among these tradesmen that was so vital to the maintenance of the Local's control over labor supply to electrical contractors. Thus the social clubs played a critical role in Local 3's ascendancy to power in New York and nationally. Based on a model of pluralism, the complex and diverse network of social clubs facilitated the brotherhood in the achievement of its goals to increase membership, continuously integrate new immigrants, keep union administrative costs at a minimum by promoting voluntarism, gather commercial information about prospective building contracts, and monitor the growth of non-union work. In addition, the clubs also provided the fledgling leadership with a base of support from which to influence politicians (especially in the city and in Albany, the state capital where electrical building codes were the purview of municipal and especially state legislators).

Van Arsdale did not challenge sex and race segregation in the labor market; instead, he adopted into the hierarchal structure of the union the same divisions that were maintained by the fraternal clubs based on jurisdiction, territory, religion, ethnicity, nationality, race, or gender. The transformative power of the clubs lies in their ability to allow the brotherhood to keep pace with economic, demographic, and technological changes occurring in the industry that threaten the de-skilling of the electrician tradesmen and potentially dilute the union's power vis-à-vis contractors. The clubs simultaneously separate and unite workers from across a broad spectrum of jurisdictions related to electrical building construction (e.g., manufacturing, street lighting, cable splicing, draftsmen, administrative support) while maintaining white tradesmen at the top and women of color at the bottom.

The effort to build union solidarity and the rise to power over New York's electrical building work was a success story. As a result, during the 1950s and 1960s the Local was able to negotiate lucrative collective bargaining agreements which resulted in the upward economic mobility of electrician tradesmen. Rising wages for tradesmen meant that these heretofore blue-collar workers could achieve the American dream and a place among the mainstream American middle class. As a sign of this, many electrical tradesmen moved from inner-city locales and neighborhoods to outlying areas such as Staten Island, Long Island, Westchester County, and other parts of upstate New York, as well as to the suburbs of New Jersey.[17]

Social Club Structure and Division Hierarchy

In Van Arsdale's view, the social club structure in the Local (based on the mutual voluntary association of earlier guilds and workers' societies) could—and did—become venues to conduct the business of a more geographically dispersed and growing union membership without increasing the overhead cost to the union.[18] These social clubs provided workers with cultural links that supplanted the kinship networks dispersed by the rise of suburbia and the flight from American urban centers.[19] Van Arsdale himself observed, "Clubs are part of a working man's society."[20] Some clubs that originally evolved out of local drinking spots in New York City neighborhoods capitalized on class solidarity and gender-ethnic bonding. "Tradesmen would walk across the Brooklyn Bridge to save a nickel. It was with that nickel that they used to be able to get a tall beer in a bar which then gave them free access to the lunch counter. There were saloons at this time in New York that catered exclusively to working men within their neighborhoods. During strikes or unemployment, these saloons would put food out for workers. So to have that nickel to get that beer to have access to that free lunch counter was a very important part of a man's carrying forward."[21] Van Arsdale continued, "Although the saloon [culture] contributed to men becoming drunk, it had, in the early days, a distinct function in the community. It was part of a working man's world. He had a club to go to where he was welcomed because he would not normally be welcomed in the regular gentleman's or business clubs."[22]

Each club has its own president and secretary who report to business agents within their division in Local 3. Members can join either on a voluntary basis or be sponsored, but as one journeyman relayed in an interview, "being sponsored doesn't hurt." Business agents play the role of power broker between club members and the central administration of the JIB. This is how such a large union with so many labor categories can sustain sophisticated operations with only a small paid staff. Each division in the Local is assigned a business agent who interacts with the elected shop stewards. The agents, along with Local 3 foremen, belong to the union's Executive Board (EB). The EB's chair selects an executive committee. The JIB and the EB and its executive committee make the policies that guide apprenticeship training, contract negotiations, pension plan administration, and other business.[23] All these entities work together to negotiate with contractor associations, hear disciplinary cases for apprentices, and survey work to enforce the industry agreement.[24] The union and electrical business owners also operate a Joint Apprenticeship Committee (JAC), composed of a director of apprenticeship, two large-scale contractors, an apprentice representative from the Apprentice Advisory Association (the apprentice club), and the Assistant to the Chairman of the JIB.[25] As of 2009, every member of each board, committee, and council is a man; there is not a single woman among them.

Of the over thirty occupationally homogeneous divisions in Local 3,[26] the Q Division is the largest. It covers craft goods manufacturing and is composed primarily of immigrant and minority women. The Local's divisions have always been autonomous and generally they meet locally. The union avoids large mass meetings of all its members, although on ceremonial occasions such as award dinners for pioneer members like the "Honor Scroll Night," men and women gather from across the divisions.[27]

Each division has a chairperson, an advisory board, and a secretary.[28] The division's business agent is its liaison to the central administration. Division officers and agents also enforce the unwritten code that obligates members across the divisions to walk picket lines or volunteer their personal time during strikes or mass organizing drives. Members' attendance at division meetings is prioritized by the local's leadership; in the construction division, for example, members are fined if they do not attend meetings.[29]

Division A—construction industry craftsmen—rules the union's roost. Business agents for each division are assigned from Division A and are expected to take the lower-paid workers under their wing. Division A workers and highly skilled workers in other divisions are expected to act as models of union loyalty. That paternalism is seen as the foundation of the brotherhood's solidarity and loyalty and is an important guarantor of the union's status quo. To the extent that solidarity implies equality, however, it includes only the members of Division A. Key figures in the Local—shop stewards, foremen, advisory board members, secretaries, club presidents—are charged with developing a loyal interest in the union among members.[30] The leadership views men in these key positions as what scholar Noam Chomsky terms the "ideological managers" of the union, and they are the main conduit for disseminating the leadership's definition of brotherhood.[31]

The construction division does include lower status workers. Job classifications within Local 3 are mind-bogglingly complex owing to the demands of a fluctuating local labor construction market, as well as to sudden or unpredictable calls for skilled labor caused by unanticipated events such as the September 11 attacks. Such considerations compel the union to intermittently devise classifications, or extend already existing ones, in order to deliver the lowest cost and most highly skilled worker to electrical contractors. The construction apprentice holds the lowest position in the construction division in theory. But the construction division M groups—trainees from nonunion sites who are mostly minority, immigrant, or women of color construction workers—usually get the most dangerous jobs.[32] As one journeyman describes the plight of the M worker, "The M group workers are called 'M dogs'." They are the lowest group of workers, below even the newly inducted apprentices (A workers). Usually, the M worker has been let into the union through either an organizing drive or government-sponsored affirmative action training program and is a male immigrant or black or Hispanic—or a woman of color. Most of the M division immigrants speak little or no English. They will spend four years as a helper, then three or four years on the M rate, then three years as a Division M worker. After eleven years, they can turn into the equivalent status of an A worker (i.e., a newly inducted apprentice), but it is entirely possible that they will never achieve their Division A journeyman card. Generally, they are an abused group of workers, given all the dangerous assignments. M group workers make comparable salaries to Division A apprentices but take a longer time (or may never) achieve a Division A journeyman's card: eleven years rather than six or seven years.[33] Most importantly, they don't have access to the social clubs. They are usually regarded, as one journeyman said, as "lower than whale shit."[34]

The Military Intermediary Journeyman (MIJ) worker is a fifth-year Division A apprentice who is placed in a supervisory position but who earns a substandard wage. This designation was devised in 1983 and in actuality serves the purpose of supplying journey-level workers to the contractor at lower wage rates.

The Local's large size and all this occupational heterogeneity make personal communication between leaders and the rank and file difficult. The Local's newspaper, *Electrical Union World* (*EUW*), is the main organ of communication among divisions. Personal stories about members' successes, failures, and problems have been a central feature in the paper since 1940, when individual shop newsletters were supplanted by the *EUW*.[35]

The Local recognizes the unequal status of workers in its various divisions by assessing higher dues from Division A workers than from workers in other divisions.[36] Dues from (mostly male) Division A workers help provide benefits for the lower-paid workers in other divisions, mostly female. As mentioned in Chapter 1, this paternal mentality regarding the organizing of women has always been at the heart of the Local's protectionism, similar to practices in craft union brotherhoods generally. First- and second-wave feminists have bitterly debated labor feminists about this paternalism.[37] As one journeywoman remarked, "This is not necessarily evil in itself, of course, but it does drive the way that men relate to women inside the union. That is, women are regarded as a weaker sorority without fully-fledged rights to power and decision-making."[38]

When men are confronted with women who want equality with them, this asymmetrical relationship is disturbed. Women as equals, making the same amount of money, doing the same physical work—this is not generally acceptable to men in these professions. It challenges their manhood, their livelihood, and their social role as chief breadwinner in the family, all at once.

Apprenticeship

The union, a brotherhood of unequals ruled by the elite electrical construction workers in Division A, is supported by the apprenticeship system. The system may maintain the hard-won economic security of electrical workers, but, like nearly every other formal and informal structure within the union, it also severely limits the mobility of women.

Apprenticeship goes back to the mediaeval guilds, when a journeyman or master mechanic passed along the knowledge of his craft to his son or other young man bound to service for a term of years.[39] But the concept of apprenticeship is more than a mere atavistic cultural holdover from days of yore; it is at the heart of the construction union's control of the labor supply for contractors—it is the union's economic power.

Many characteristics of apprenticeship training from the time of the guilds survive in today's construction apprenticeship programs. For instance, the concept of apprenticeship embraces not only the transmission of skill but also pride in craftsmanship and the manly status brought by earning a family wage.[40] Especially today, when mental labor is more highly regarded than manual labor, these cultural aspects of apprenticeship in Local 3 remain a vital part of the concept of brotherhood.

The transmission of skill through apprenticeship has important gender implications. Traditionally, the relationship of master craftsman to apprentice has been that of a man to a child, a father to a son. The acquisition of a craft skill and the maturity to support a family are synonymous with manly pride and masculinity.[41] Apprenticeship, as defined in Local 3, is yet another form of male solidarity where the apprentice "kid" not only acquires the skill but also the wisdom of the journeyman, at work and in the family. It is difficult to integrate women into this mentoring process.

In most industries, the labor supply is controlled through the mechanism of the market: when employers need more workers, they usually raise wages. When demand for the product or service falls, employers need fewer workers and can unilaterally lower wages. In the construction industry and some others, however, this process is mediated by the unions. The union bands workers together to bargain collectively with the employer, resulting in wages and working conditions that are subject to ongoing negotiation rather than to arbitrary change.[42]

In many industries, workers are hired regardless of their membership in a union, although they may be required to join or contribute to a union to keep the job. Construction is different. In unionized construction, a worker has to have been, for all practical purposes, "hired" by the union (i.e., to have a union card although the actual "ticket" for a job is issued formally by the JIB) before he or she has a chance to bid for work from employers.[43] The apprenticeship system is the mechanism by which construction unions attempt to control the labor supply by contractually establishing an agreed upon ratio of apprentices to journeymen on work sites.[44]

Employers claim the unionized apprenticeship system serves little purpose except to inflate construction costs. They argue that only union power and coercion maintain this system and employers can hire as-skilled but much cheaper labor from the nonunion ranks.[45] A 1947 article from *Fortune* magazine, "The Industry Capitalism Forgot," gives a more accurate assessment. Construction, it explains, is "the one great sector of modern society that has remained largely unaffected by the industrial revolution."[46] Its labor force is simply not amenable to the canons of "scientific management," let alone advanced automation. A skilled tradesman cannot adequately be replaced by an unskilled one or a machine. The men and women hired to put up a building from blueprints must know how to use tools and how to work safely. Only the union apprenticeship system adequately prepares young construction workers for their jobs.[47]

This educational function and the notion of skill in general are vital to a union member's image. Electricians develop their own work styles but all display pride in their abilities to maintain a certain degree of skill in the trade, known in the conventional wisdom as "the standard."[48] They compete with each other to show off their skills at installing panels or using power tools. The imposing skyscrapers that glitter against the night sky testify to those skills. As one male journeyman explained: "Since the early nineteen teens, we've been there lighting the history of this incredible city—I'm so proud in what we've been able to accomplish—the Woolworth Building, the Chrysler Building, the Trump Building, the GE Building, the Financial Center, the newly planned Freedom Tower to replace the Twin Towers at the World Trade

Center—and I could go on forever. These are our version of modern day pyramids and the electricians' skill glows there in all its glory."[49]

In the brotherhood, respect for the highly accomplished electrician supersedes politics and connections. The union holds that if you learn well, you will succeed. Young male apprentices are advised that their job security will depend on "how well you learn the trade."[50] The apprentices are trained to be skilled artisans and also loyal team members.

At times employers prefer unionized apprentices over higher-costing journeymen for financial reasons. Employers may also prefer unionized apprentices over nonunion workers due to the fact that the former have had on-the-job training supervision and an exposure to the broad mastery of electrical work, whereas nonunion may not. This preference also depends on whether or not highly skilled immigrant labor (often nonunion) is readily available to employers who may have to hurriedly scramble for thousands of workers across the trades to staff large-scale, federally funded projects. A ready supply of accessible and predictable unionized apprentices and journeymen made available by the union and joint board allows employers the flexibility to shorten the proposed length of time (and the cost) of large-scale federal projects, thereby making unionized employers more competitive than nonunion in the contract bidding process for federally and state subsidized work and more likely to secure projects. The union's ability to guarantee a predictable number of skilled workers to a unionized contractor consequently strengthens their position in the industry. The union is naturally reluctant to provide too many apprentices, in fear that apprentices will eventually displace the demand for journeymen; but it can provide more of these cheap—yet skilled—workers in return for concessions from the employer.[51]

Erratic Employment

The iconic phrase "the personal is political," coined by Sara Evans in the 1970s to describe the day-to-day impact of larger societal economic and political forces on the quality of individuals' everyday lives, is relevant to the way in which social relations on the job and in the home are shaped by tradesmen's job insecurity due to the volatility of the construction industry.[52] I remember the terrible and constant feeling of economic insecurity of my father's employment in the trades. It affected all the relationships inside our family and contributed to an ongoing tension and fear. I believe my father always tried to compensate for this by his overbearing and macho attitude toward my mother and his children. Perhaps part of the enduring discrimination toward women in the industry stems in part from this insecurity which undermines a man's sense of identity. Employment in construction is more erratic than in almost any other occupation. Inclement weather, seasonal unemployment, housing market slowdowns, and economic recessions are the perennial realities of the industry. During the current economic recession, unions are making concessions to builders that will lower building costs in order to avoid sharp unemployment surges among unionized workers in the industry. However, during the recession of 1972, more than one million building trades workers, nearly 25 percent of the total industry workforce, were unemployed. But it was the union-controlled joint industry board, not the employers, who decided who worked and who did not through union seniority systems and the demand of contractors for certain skill levels. The power of the building trades unions to control the supply of labor—and with it the power to smooth out the pain of erratic demand for skilled tradesmen's services—accounts for their prominent position in the labor movement.

In general, this union stewardship of the labor supply, with its ability to bypass the cruel calculus of laissez-faire economics, has been profoundly beneficial for blue-collar workers. Of course, the stewards may not always be fair. They can determine the populations from which apprentice applicants will be drawn—which ethnic groups, which races, which gender. By 1900, most journeymen electricians were men of Northern European extraction—English, Dutch, Irish, and Scandinavian.[53] Periodic scarcity of work in the trade led these electricians to exclude most newcomers other than their own sons or other relatives. They wanted to keep work "in the family." Family meant loyalty, and loyalty meant greater union discipline: if apprentices are disruptive on work sites, then it slows down the flow of work, delays other trades' work, costs the contractor extra money in time and materials, and may even jeopardize the building contract deadline which, in turn, will mean higher interest rate costs on loans to the contractor. The contractor may even decide to drop the union workers and hire nonunion scabs.

Union discipline as a value transmitted across generations in the trade has led to stronger bargaining positions against recalcitrant employers and greater control over job jurisdictions and hiring for the Local. Integrating whole families into the trades gives the union greater control over the discipline of young male apprentices on job sites. If the young men drink, fight, or just become unproductive "stiffs," their families will find out.

Restricting membership in the trade, of course, eventually enhances the journeyman's own market worth in the short run. Ultimately, the only mechanism the union possesses to sustain this discipline and inspire loyalty among members is the provision of continuous employment.

Leadership positions in the union are also handed down like a family heirloom from one generation to another. So in the 1980s, when Harry Van Arsdale left the office of IBEW business manager, in keeping with the "father-to-son" tradition, the post went to his son Thomas, and at the beginning of the twenty-first century to his grandson, Christopher, a very practical recognition of Local 3's oligarchic structure.

Birth of the Modern Apprentice System

Prior to the Great Depression, electrical contractors sent construction helpers ("permanent boys" in the jargon of the trade) to a four-year training program. Instruction was provided by union journeymen. The helpers paid a fee to the union, and the Electrical Contractors Association paid the helpers for the time spent in school. When the helper completed his training he became a journeyman in the union. The 1929 stock market crash curtailed this practice.[54]

In 1937, the New York State Legislature passed regulations for oversight of apprentice training programs as mandated by the federal National Apprenticeship Act, conventionally known as the Fitzgerald Act, which established the first state-supervised training program in the building trades, subsidized by state and federal funds. The Local subsequently developed its own apprenticeship program under compliance with state regulations and the union replaced the term "helper" with "apprentice." It was also not long before the wartime shortage of young men again compelled the Local to curtail the promotion of apprentices to journeyman status, and the training program was delayed until the end of the war. Afterward, apprentice classes grew steadily in size and the apprenticeship program instituted more formal work rules, orientations, and ceremonies marking promotion and graduation. By the mid-1950s, classes and curriculum had become standardized and the JIB enforced work rules regarding absenteeism.[55] Male apprentices were sent to relatively small shops where they got real technical training, not to major construction sites where they were often relegated to unskilled manual labor. They also attended six hours of classes per week in electrical theory.[56] In addition to training new workers, the apprenticeship system introduced them to the concept of fraternity. Selective admission contributed to the cachet of the apprenticeship program: from the beginning, apprentices considered themselves as part of the elite. The elite identity of apprenticeship was enhanced in the 1960s when civil rights activists pressed the union to open up admissions to the program and father-to-son "legacy" admissions fell into disfavor.[57] Instead, candidates for apprenticeship were interviewed by the Joint Apprenticeship Committee composed of four interviewers (two each representing the employers and the Local). The candidates selected were then sent for physical examinations.[58]

In recent years the process has expanded to involve fifty interviewing teams composed of rank-and-file journeymen (recruited mainly from the Local's social clubs), shop stewards, foremen, contractors, and a representative from the state.[59] Once a year, approximately six hundred applicants are interviewed in one day at the JIB. Although there is a manual dexterity test administered by a state proctor, its total value is only 80 points; the interview itself is weighted at 500 points.[60] Successful candidates are then referred to the Local's EB to be indentured—that is, inducted into the apprenticeship program and sent out to a job site to be apprenticed to a journeyman.[61] Local 3's collective bargaining contract with the ECA calls for the ratio of one journeyman to three apprentices; this ratio is of course always subject to renegotiation as industry conditions change.[62]

By 1971, about 2,000 apprentices were available to contractors and that number held steady in the 1990s until the collective bargaining agreements as recent as 2008.[63] Entrance into the Division A apprenticeship brings with it not only a journeyman's salary of approximately $75,000 per year (after six years of on-the-job training and electrical theory classes at a New York City high school), but also a high degree of status and prestige. Generally, about 5,000 applicants per year vie for 500 apprenticeship positions. Only about half of all apprentices ever become journeymen electricians. Job dismissals, injuries on the job, a drinking culture, and other disciplinary problems all contribute to the high rate of attrition.[64]

The post-World War II reforms in the apprenticeship system proved a boon to men from groups previously excluded. The first post-war apprentices were predominantly Irish, Jewish, and Italian. The few minorities who gained access to apprenticeship did so through an aberration: mothers in the Local's manufacturing division (Division Q) were allowed to sponsor their sons as apprentices in the construction division. As the ethnic composition of the manufacturing division changed to include more African Americans and Hispanics through mother to son sponsorship, more minority men gained entry into the apprenticeship program overall.

During the building boom of the 1950s, the Local opened the Division A apprenticeship to all union members' sons who held a high school diploma and were between 18 and 21 years old.[65] In 1962, demonstrations at work sites in Harlem by black electricians (led first by A. Phillip Randolph and later by Adam Clayton Powell) forced the Local to strike a deal with the Kennedy White House concerning breaking ground for Harlem Hospital, a federally funded building project. Harry Van Arsdale promised the president that the electrical workers in New York City would be the first craft trade union to voluntarily take in a "Class of 1,000" Black and Hispanic (mainly Puerto Rican) apprentices, and serve as a role model to other construction unions.[66] This concession would help Kennedy politically with civil rights activists. In return, Kennedy promised to promote legislative efforts to eliminate the restrictions on the use of apprenticeship labor on skyscraper work in New York City (where they had been prohibited). In one fell swoop two of Van Arsdale's goals were achieved: labor costs were kept down for union contractors working on skyscrapers and union jurisdiction was preserved over a large share of electrical work in high-rise construction. The Class of 1,000 minority men were recruited primarily from the ranks of the

Local's manufacturing division through mother-to-son sponsorship. Minority male recruits were also sponsored into this first class from black advocacy organizations such as the A. Phillip Randolph Institute and its breakaway organization, the fledging Harlem Fightback, a black militant advocacy association aimed at integrating minority men into building construction work.[67]

But the union's movement toward greater racial diversity in the postwar years did nothing to break the fraternal bonds of gender solidarity. Women were simply not considered candidates for apprenticeship. The all-important Joint Apprenticeship Committee (JAC)—composed of union officials, employers, and staff members of the JIB—still counts only one woman among its twenty members. Although women have been elected to the ranks of division chairpersons and advisory board members, there are no women representing the large and mainly female manufacturing division on the Local's EB or its Executive Committee, two bodies that act as close advisors to the business manager.[68]

Discipline

Even as apprentices, women are treated differently from men. Formal disciplinary actions against male apprentices are technically the purview of the JAC. It is rare, however, that a male apprentice is brought up on formal charges before this group; the errant apprentice is usually informally advised to change his ways. Only male apprentices who are not connected to anyone in the trade go before the JAC. If a male apprentice is drinking on the job, not showing up for work, horsing around, displaying porn to other workers, or otherwise slowing production on the job site, his sponsor is asked to straighten him out and remind him that the JAC could throw him out of the program. Similarly, if a minority male was recommended by an outside organization such as the A. Philip Randolph Institute, the officers of the Local would refer the problem to the sponsoring group before bringing the apprentice up on formal charges.[69]

This informal approach to disciplinary action does not apply to women.[70] Instead, infractions such as absenteeism are immediately taken up by the JAC. No mentor investigates or coaches the female apprentice. Absenteeism among female apprentices is commonly a response to harassment on job sites, but when women are brought up before the JAC this is never taken into consideration and they are likely to be simply thrown out of the program. Evan Ruderman, one of the first female electricians in Local 3, stated:

> The way in which the Local deals with the men is totally different than women. Instead of being "unofficially" reprimanded by a foreman or friend in the industry, they are more likely to be brought

up immediately on charges and brought before the Apprenticeship Council for disciplinary action. Often times, women will use absenteeism as a way of coping with hazing, or harassment from the guys.[71]

Recently, a woman apprentice was sent to the JAC for disciplinary action. A union sister explained that "she was actually being set up, because the Director of Apprenticeship had already called down and asked to have the woman thrown out. Luckily, there was a woman on the Committee, the first woman shop steward in the Local's history, and she persuaded her male committee members to go easy on the female apprentice."[72]

Lucky indeed, since filing a grievance is rarely a good option in Local 3. Male workers do not use a formal grievance procedure. As in other building trades construction unions, job issues are usually reported by telephone to a construction desk inside the Local. As with police and mineworkers, there is a code of silence, a brotherhood-in-arms mentality among the men in Division A, where "airing dirty laundry in public" through a formal grievance jeopardizes the status of any worker, male or female.[73] Most complaints or problems concerning male workers that arise on the job site are handled informally through the network of club members, and through a male shop steward or the Joint Industry Board. Even within a formal grievance process, women routinely report getting the runaround from shop stewards and business agents when they try to file a grievance against either a male coworker or foreman or straw boss (a union member supervising a crew of about ten workers). Often, grievances by females are reportedly just dropped by the union.[74]

Apprentices and "The Estate"

Newly indentured apprentices are first gathered for orientation sessions at the headquarters of the JIB, where they are given a package called "Your Estate."[75] This package describes the apprentices' responsibilities, benefits, and future prospects in the industry. In its language, rituals, and style, the "Estate" kit conveys the strong impression that the collective they are joining is a male organization.[76] One journeyman stated, "The electrical training is supervised by journeymen who are handpicked by the Local's leadership." Volunteering as an instructor in the apprentice program elevates journeymen to a higher level of respect in the brotherhood. And the sacred trust of passing knowledge from journeyman to apprentice is not only technical but also political.[77]

The "Estate" pamphlet uses language derived from nineteenth-century fraternities to describe an inheritance that is imbued with all the rights and obligations of any inheritance. For instance, the estate should not be "squandered" or "multiplied."[78] Once admitted into Division A, apprentices become "indentured."[79] This indenture makes these young men eligible to get work tickets at the JIB for job assignments. "But make no mistake," a former director of apprenticeship, Buddy Jackson, said, "It's also a way of teaching these young men about being a man and what that means at home and at work, as well as the rules of the union and trade."[80]

During orientation, apprentices get a list of tools and clothing they will need for the job, a set of basic safety instructions, and a work ticket from the Employment Office. At the completion of a job or after a dismissal, apprentices are laid off. A good layoff means that the job was completed and the apprentice may return to the Employment Office for a new job assignment. A bad layoff is the result of poor performance, excessive absence, or other undesirable behavior. A bad layoff brings a hearing before the Apprentice Termination Review Committee. The apprentice is usually placed on probation and given another chance. Foremen evaluate apprentices in their shops every three months. Both foremen and journeymen are charged with the responsibility to teach apprentices so they will be competent by the time they reach journey worker status. Apprentices also take classes in the electrical trade and labor studies. The required courses in electrical theory are taught by journeymen, and as stated earlier, this teaching is highly valued as a form of union activism.

Apprenticeship teaches new workers the formal and informal work rules that govern the brotherhood, and helps them take pride in their craft. Much of this learning takes place on the job, often through a buddy system pairing younger and older workers. This indoctrination into Local 3's fraternal union culture is also reinforced in sports teams, junior fraternities such as the apprentice club, and especially in fraternal social clubs.[81]

The Many Roles of Social Clubs

Van Arsdale was well aware of the ethnic and racial diversity in New York City, and understood why union solidarity had to cross these divisions. Under his leadership, the Local's sponsorship of fraternal social clubs, some based on ethnic and national identity, helped the Local's elite deal with the ethnic layering by attempting to build solidarity across cultural differences.[82] These clubs also helped knit together a membership dispersed throughout the boroughs of New York City.[83]

Although Local 3 leaders insist that any of its thirty-five thousand members can join any club, members do have to be sponsored, albeit in an unofficial way, to get into many of the more powerful clubs in the union, and club membership is semi-exclusive. There are currently twenty-four social clubs drawing members from across the divisions.[84] In addition to one club in every borough, there are clubs for members of each major religion, ethnicity, and nationality represented in the union. There is also a club for women members, a club for retirees, and a club for foremen.[85]

In these clubs, "a fraternal feeling and a spirit of lodgism are strong."[86] Local 3's highly evolved club structure is unusual. The clubs serve as nerve centers of the union's affiliation network. According to labor historian Alice H. Cook, although attendance at union meetings is high, they are not the only place where the rank and file are expected to participate. Local 3 is bound together by a network of social institutions that fill the leisure time and satisfy the social needs of many of its members, but particularly those of its pacesetting Division A.[87]

The designation "social club" is actually something of a misnomer. The social clubs in the manufacturing division are just that—social clubs. But in Division A, the clubs function as a structural part of their antecedent union administration. The Local vehemently denies this; every Local staff person interviewed maintained that the clubs are merely voluntary organizations where workers "chit-chat" and can gather and exchange information or seek recreation and leisure. Interviews with rank-and-file members, however, contradict such assertions.

Despite the fact that social clubs are in theory open to members from all divisions of the Local, it is invariably Division A construction workers who are most prominent in the clubs, especially those that are older and more closely linked to the union's power structure. Once a member receives a journeyman's card in Local 3, he or she has access to the inner circles of the brotherhood in one of the more prestigious social clubs in Division A.

During the apprenticeship, however, the member can volunteer for union activities and network with influential foremen and stewards in the Local by joining the Apprentice Advisory Committee (AAC), the junior fraternity of electricians in Division A. This too is an apprenticeship of sorts. The officers of the AAC, all apprentices, are chosen for their potential to loyally serve Local 3's leadership, which means not raising any questions about the policies or practices of either the union or the JIB. In other words, union discipline equals loyalty and equals brotherhood. So although the club encourages a focus on athletic activities (competing on the junior fraternity's union-sponsored softball team) and philanthropic work (sponsoring a Christmas party for poor children and raising money to buy gifts for children living in foster homes), membership in the AAC is hardly all about good works and recreation.

In addition to learning about the job and the rewards of union discipline, male apprentices are kept in line by their male mentors in the AAC. Dissidents are quickly weeded out and more accommodating apprentices are promoted to officer positions in the club and—once they achieve journeyman status—sponsored for membership in one of the top social clubs, such as the Allied Social Club, the Catholic Council, or the Westchester Mechanics.[88] Though each of these influential clubs has only a few officers, club members can network within the top-tier circles of the brotherhood. This often leads to better jobs, overtime pay, and advance information regarding available work. Local 3 does not have a similar path that is effective for women who aspire to leadership positions.

Like apprenticeship, clubs have their roots in the benevolent and cultural craft societies of the nineteenth century. Each club has its own business agent, a paid staff member from the Local who reports to top officials and plays the role of power broker between club members and the central union administration. These fraternities function on more than just a recreational level; they are more than mere "clubs." Organized under union auspices, they are part of an unofficial governance structure based on male gender solidarity. This governance structure historically has tended to include women only as auxiliary, albeit important, members. In Local 3, women union members in the manufacturing divisions are auxiliary clubbers, and excluded from full participation, at least in clubs that are central to the governance of the Local.

The importance of the social clubs to Local 3's cohesion, order, and internal discipline cannot be overstated. Reportedly, clubs compete for well-placed members in order to strengthen their own power base. Clubs that can recruit officers of the Local, for example, have far greater access to information concerning the availability of choice work assignments and overtime. In turn, well-connected and active clubs enhance the power and control of the Local over all the membership. Business agents and other officers are always in close contact with the more important clubs. While the Local exacts a tribute of service from these clubs, in return it supports the clubs by hosting meetings on site at the JIB, providing hospitality, assigning paid staff business agents to resolve issues for club members, and providing opportunities for members to advance on the job and within the larger union structure.

Members may be sponsored to join especially desirable clubs—or discouraged from ever trying. This serves a disciplinary role. For apprentice and journeyman alike, clubs teach a discourse of acceptability by making it clear what individual members may and may not speak about in public. They "make workers one-sided so they cannot see what is going on around them," complains a male journeyman.[89] Potential shop stewards, business agents, foremen, and straw bosses are all drawn from the ranks of the clubs, which serve as screening mechanisms and promotional ladders. Clubs often function as self-help organizations and support the networking so necessary for advancement in the union. Members who are willing to organize dances, picnics, weekends away, award dinners, and various volunteer community projects become buddies. The club structure makes it possible for a journeyman, even one who was not brought into the union by his father, to attain status and position—even a leadership role—through voluntary work. In return for their service, the union places them in an elite circle where they are less likely to be fired or laid off. As one journeyman puts it, if you want to get places within the union, "the social clubs are the place to be."[90]

One social club even helps chart the union's future. Confronted by the city's changing demography and rapid technological change, Local 3 established its Futurian Society in 1960 to study and advise on these issues for the business manager.[91] Its members are exclusively college graduates. Rank-and-filers call them the "explorers" of the union; Van Arsdale called them the "elite of the elite," Local 3's own in-house think tank.[92] In 1975, one of the worst years for construction work, the Futurian Society was directed to establish an international information center to monitor the availability and training of craft workers in Mexico and Europe.[93] Members kept a close watch on the exporting of jobs and the encroachment of cheaper craft labor through both legal and illegal immigration. In the face of vast socioeconomic and political transformations in the larger society, social clubs like the Futurian Society are an institutionalized means of coping with diversity, controlling the expansion of nonunion work, and encouraging volunteerism to strengthen bonds among the union membership.[94]

Alice H. Cook described the principal function of these organizations through the analogy of precinct clubs in political organizations: "They are centers where members meet socially, run their affairs, and carry on in the hours after work the shop talk, camaraderie, and friendships established on the job."[95] In concrete terms, these social clubs engage in much the same activities as organizations such as the Elks: running dinners, picnics, athletic contests, and the like.[96] In his study of Local 3's organizational behavior, Cornell professor of industrial and labor relations Maurice F. Neufeld finds especially striking the clubs' many philanthropic activities, like raising funds for charity and sponsoring gendered fraternal organizations, such as Boy Scout troops.[97] The union makes considerable effort to provide both male-only activities and functions for members and their families.[98]

Beyond the function of socialization into Local 3 culture, the clubs help support the status quo within the Local in three other ways. First, they serve as training grounds where well-connected journeymen can be identified and cultivated for leadership positions. For instance, Local 3 journeymen who are not attached to a specific club and who have never served in an official capacity within one of these clubs rarely if ever attain union office. Second, the clubs provide an outlet for union activism which is safely directed away from the core decision-making process. In effect, according to many rank-and-filers, the clubs are a diversionary mechanism by which active union members are discouraged from questioning union policy and the tenure of incumbent union officials, and instead devote energy to benign and approved social activities. Third, union social clubs slowly integrate new groups into the union's power hierarchy. Over the decades, new ethnicities such as the Jews and Italians (Electrical Welfare Club and Bedsole Club, respectively) entering the trade—most recently Asians and Greeks—have been encouraged to form their own clubs. These clubs, such as the Asian American

Cultural Society (mainly Chinese) and the predominantly Greek St. George Association, are incorporated into the social club hierarchy at the very bottom of the power ladder, and the constituent ethnicity only slowly integrated into the union power structure.[99] As these clubs became more powerful, other ethnic clubs replace them on the bottom of the ladder. It is a de facto form of ethnic layering on the job—a clean form of segregation. The ethnicity of every worker is at once recognized as a way of building and sustaining solidarity in the union, while at the same time dividing the membership along the lines of ethnicity and nationality through this semiformal organizational matrix.[100]

Meanwhile this fraternal culture of the union's social clubs makes it difficult for women to become involved and influences how long they stay. At first, the women of Local 3 had no club at all, only a Ladies' Auxiliary in the manufacturing divisions designed to organize annual social outings for the men and their families and to assist union leadership in running union elections in the boroughs.[101] When the pioneering women electrical construction workers did form a women's club, the Women Electricians (WE) of Local 3, they debated over whether to affiliate it formally with the union, and thus acquiesce to the union's power structure, or to make it an independent group, thus denying it any voice within the union's corridors of power. The debate threatened to render the club impotent, forfeiting any advantage of either course of action. Nonetheless, the Local's overarching opposition to sanction the club or to be responsive to WE's members far outweighed any negative consequences of internal divisions among women members. Consequently, the WE leadership felt hamstrung in its ability to effectively influence the union's policies toward greater gender equality for women construction workers. In the meanwhile, the Local co-opted some activists and former members of WE and established a women's club that was controlled by the male leadership.

Electchester

Years ago, meat cutters had Concourse Village in the Bronx, garment workers had Seward Park Houses in Manhattan, and printers had Big Six Towers in Queens. Various local unions sponsored more than a dozen cooperatives that provided almost 50,000 decent and affordable apartments to working-class families. As the power of labor waned and real estate values skyrocketed, though, most of these co-ops either lost their direct union connection or allowed their shareholders to sell their units on the open market. These days, the only co-op left with strong union ties is Electchester in Flushing, Queens—the pride of Local 3. In fact, Local 3 was the first union to undertake a housing project for its members, and this kind of building project was unique among the skilled building trades unions.[102]

The Local has always sought to strengthen the bonds of union solidarity by strengthening the nuclear family and providing community support. In 1949, Harry Van Arsdale Jr. worked with the JIB to purchase a 103-acre former country club in Fresh Meadows, Queens.[103] The completion of Electchester with its first housing unit completed in December 1951 (there were to be twenty-four more completed by the 1960s) is a monument to the union's determination to create community and solidarity among its members. The apartments were built by volunteer labor from each division and social club in the Local, who worked on the project after regular hours and on Saturdays. The project took shape at a time when there was very little other residential construction in New York City, but today the Local struggles to maintain that extended-family intimacy among unionized electricians.[104] Not surprisingly, some Electchester apartments contain complicated wiring schemes, due to the predilections of shareholders to upgrade the wiring themselves.[105] All told, 5,550 people live in about 2,500 units in thirty-eight buildings, many of which are six-story structures.[106] There are plenty of trees and playgrounds, a public school nearby, and the complex is a short walk from the Electchester shopping center.

A neon sign marks the entrance to the Electchester complex, which includes an array of small shops and restaurants. Electricians' wives wheel strollers down Harry Van Arsdale Jr. Avenue. "Harry Van did a beautiful thing, because it was very good for the average individual," said Leo Sprinter, a spry man in his early eighties. Sprinter worked for the Army Corps of Engineers and has lived at Electchester since 1953. "Even as an outsider, I have a certain feeling for Local 3, because they're all around me."[107] People often gather near the imposing JIB Center. Electchester residents old and young come here for lunch, shopping, or conversation over coffee or pizza. Plastered on every bulletin board inside the shops are notices of upcoming cultural events, testimonial dinners, and social or community service activities sponsored by Local 3 and the JIB. These events include picnics, seminars at Camp Solidarity, Boy Scout events, family days, dinner-dances, bowling leagues, ladies' nights, and plays and concerts.

The JIB Center was built in 1961 to complement the community established by the cooperative housing project. The Center includes a bowling alley.[108] Here, many electricians' wives spend a few recreational hours. The "JIB Lanes" provide a coffee shop and on-site day care for Electchester residents. Although union insensitivity to child care has been a major source of contention for women electricians, for wives ensconced safely at home, the union does provide limited child care. The bowling lanes are also frequented by retirees' associations from both Division A and the electrical manufacturing divisions.[109]

Electchester also houses a huge auditorium where apprentice union meetings are held on the first and third Wednesday of every month.[110] On other nights, the social clubs sponsor various community and cultural events.[111] The fraternal social clubs use the bowling lanes on Friday evenings and one other time each month when all the clubs meet for a tournament, an opportunity to socialize across fraternal organizations.[112]

The union electrical journeyman is one of the rare blue-collar workers whose wages allow a single breadwinner to maintain a nuclear family. This privilege can have a price. Waiting for a cup of coffee in a booth in the Electchester coffee shop, I overheard what could have easily been a scene out of "Desperate Housewives"—two wives of electricians whose children are in the JIB daycare center across the street. Over lunch, the women discuss some of the intimate details of their lives with their electrician husbands and problems of infidelity. "I can't stand his escapades anymore," says one woman. "But do I have a choice? He makes so much money, the kids and I really need him."[113]

Here in Electchester the dream of the union as central to work, community, and personal life has come true.[114] It is also a model of unionism with deep roots in the nineteenth-century labor movement. Other unions have tried to create this kind of a community—the Service Employees International Union (SEIU), Local 1199, the former International Ladies Garment Workers Union (ILGWU, now merged into the United Needle Industry Trades Union, UNITE!), to name a few.[115] None have been as extensive or successful as IBEW Local 3. The hiring hall, the shopping mall for clothing, groceries, toys, and the huge adjacent housing complex for electricians: everything is here in one community.

Over the years some electricians have moved their families out to the suburbs, and the proportion of union members living in Electchester has declined. Recently, some longtime residents have expressed interest in cashing out their shares on the city's tight housing market.[116] But the Local maintains its commitment to supporting the intimacy of an extended communal family among union electricians. This determination likewise continues to be a blatant demonstration of the union's patriarchal attitudes toward women. Predictably, when I interviewed a union official, he claimed that the union has never discriminated against women. He said, "It's merely a matter of knowing one's place. . . . Women don't belong in the industry—although they have every right to be in the workforce if they choose to do so."[117] Electchester is the "place" for family and community life and the actual "place" an electrician's wife should know where to be.[118]

Fraternalism is embedded in Local 3's political work. Local 3 often asserted the power of community and solidarity in the realm of electoral politics, in the great New York tradition of fusion politics.[119] In 1962 the union established a political party, the Brotherhood Alliance, using the volunteer labor of male union members from across all religious, ethnic, racial, and religious social clubs, including the predominantly African American Transport Workers' Union.[120] Transcending racial divisions, this one-issue coalition staged a massive voter registration drive and canvassing effort, carried out through the social clubs of Division A. The Local enlisted the help of civil rights activists and black trade unionists, even inviting Dr. Martin Luther King Jr. to its opening meeting.[121] According to *Electrical World News*, "Dr. King wished the [Brotherhood Alliance] success and pointed out that 'What labor needs, Negroes need—we must live side by side.'"[122] The Brotherhood Alliance organized sixty-five assembly districts and staffed the offices with pioneer members primarily from social clubs and from all divisions of the Local.[123] Van Arsdale headed the Alliance's governing board, which included representatives from other craft unions, as well as building service and utility workers union officials.

The Alliance refused to align itself with any one political party during this period and emphasized its independence, supporting Republican John Lindsay for mayor in 1965 as well as another Republican for borough president. "We are our own party," an Alliance representative proudly told the *Electrical Union World*, "and we are committed to the proposition that now and in the future we will select our candidates solely and exclusively on the basis of their ability and integrity, regardless of party label."[124] Such a strong and independent political commitment at this time in the city's history was unprecedented for an AFL craft union, and was a bold attempt to identify the interests of craft workers with the interests of New York City itself.[125] Van Arsdale professed to see no conflict "between the aims and aspirations of working people and their families and all others who live in this city."[126]

For all its lofty promises and fanfare, the Alliance was disbanded soon after the election, having served its purpose: to demonstrate the power of labor and to elect Robert Wagner mayor. Its alliance with the civil rights movement vanished with it.

Like apprenticeship and the social clubs, Local 3's foray into electoral politics cannot be seen as merely ancillary to its union goals, but should be viewed as essential to promoting a culture of brotherhood—the concept of the mutual duties owed by the union to its members, by the members to the union and to one another, and by the members to their families. Like the Local's softball tournaments, the Futurian Society think tank, the daycare center at the bowling alley, the electrical theory classes given at the JIB, and the great housing and recreational complex of Electchester, the Brotherhood Alliance was one more way the union reinforced its message of voluntary solidarity, supported by gendered ideology on and off the job. Ideally, the union member and every member of his family are bound up in an organic, functioning whole, a united front against any who would challenge their role in shaping life on the construction site.

3

The Struggle to Become Electricians

My older brothers and sisters were involved in the 1960s in the Young Lords, which was like the Puerto Rican Black Panthers. I used to wear a black leather jacket. My older sister and brother understood more what was going on. But I knew the jacket meant the liberation of the people from racist oppression. Pablo Guzman was in Brooklyn. And I was affected by what was going on; although I was young and by the time I got to my generation, it was the feminist generation.[1]
—MELINDA HERNANDEZ, PIONEER ELECTRICIAN

Women's struggles to enter the electrical brotherhood have been shaped not only by the Local's traditional fraternal structure, but also by factors such as the historical contingency and influence of national and local public policy, the feminist movement, and community grassroots activism. These factors all challenged the formidable gender and racial hierarchy of the construction industry and the electrical brotherhood.

The National Story

Modern scholars of women's labor history have described how a gendered social order penetrates all institutional aspects of American society.[2] Workers and bosses, labor leaders and capitalists alike, have agreed about one thing: a woman's place was in the home—or, at the very least, in a female-typed occupation, such as nursing or teaching.[3]

During the Second World War, labor unions in auto manufacturing and many other industries supported women's seniority rights and their right to earn wages equal to men's. But when men began returning from the war, manufacturing work again became "men's jobs." Laws stipulating that returning veterans had a right to their previous jobs meant that women who had been "filling in" were now displaced. Employers who refused to restore veterans to their previous positions broke the law and risked public condemnation.

So in such traditionally male and highly remunerated work as construction and welding, women were expected to give up their jobs to returning veterans and return home voluntarily. "Absent any sense of permanent change in the roles of men and women, women's individual desires remained secondary to a shared commitment to families," writes women's labor historian Alice Kessler-Harris.[4]

Not every woman wanted to give up her job, a job at which she had become adept and which gave her a more powerful self-image and independent income.[5] Black women and single mothers did not embrace the same shared commitment to family as married white women who saw returning to their domestic roles as the best or only option. Indeed, their role as breadwinners motivated them to fight vigorously for their "Rosie the Riveter" jobs, which paid a family wage.[6] Sociologist Ruth Milkman relates what happened when Ford Motor Company managers dismissed women workers with seniority in 1945. The women appealed to officials from the United Auto Workers (UAW), who initially refused even to meet with them.[7] Mildred Jeffreys, who later that year became a UAW vice president, remembered, "The message we got from the Board was: Just who do you women think you are, anyway?"[8] The union brothers shrugged at their sisters' outrage. The women angrily appealed again to the UAW leadership, and the union leaders then changed their position. They explained to the male rank and file that if Ford could "get away with disregarding seniority rights of women workers now; they will be in a stronger position to disregard seniority rights of other workers later on." Therefore, protecting the rights of women workers would make it easier for male unionists to defend themselves later.[9]

The UAW's support for the seniority rights of wartime women workers was an all-too-isolated event. After the war, and in collusion with employers, union leaders hung a "men's only" sign on skilled jobs. It remains in place today. Milkman has demonstrated that defining jobs as "male" and "female"[10] is complex and historically contingent. An industry's formative period is critical: once a pattern of employment by sex has been established, it soon becomes inflexible. Employers do not challenge it, even though they might lower labor costs by hiring women. Men and women alike accept the divisions as, simply, "natural."[11]

The Civil Rights Framework

The Civil Rights Act of 1964 was conceived, drafted, and debated as a measure to protect the civil rights of black men. Over the protests of black leaders, it was drafted to protect the rights of women as well. Title VII of the Act prohibits employers and unions from discriminating based on race, religion, national origin, or sex. "Sex" was added by Virginia Congressman Howard W. Smith, who chaired the House Rules Committee and led the conservative coalition that opposed the Civil Rights Act.[12] Smith offered his amendment two days before the congressional vote on Title VII. According to Smith's colleague, Congressman Carl Elliott of Alabama, "Smith didn't give a damn about women's rights . . . he was trying to knock off votes either then or down the line because there was always a hard core of men who didn't favor women's rights."[13]

The linkage of sex to race, religion, and ethnicity in the Civil Rights Act could never be called comfortable. It was viewed as a way of killing or weakening the legislation. As historian Nancy MacLean has pointed out, the Equal Employment Opportunity Commission (EEOC) fumbled and stalled on its way to achieving any kind of parity in the treatment of race and sex in terms of the Act.[14] The EEOC did not conclude until 1969 that sex could be treated like race for administrative purposes.[15]

Under tremendous pressure from civil rights groups, one month after the Act went into effect President Lyndon Johnson issued Executive Order 11246, prohibiting racial discrimination by federal government contractors and mandating that employers take preferential action to correct statistical disparities between the pool of minority workers available for jobs and the numbers actually employed.[16] The executive order said nothing about women.[17] But chastised by Esther Peterson, a leading activist in the labor and women's movements for more than forty years, and under intense pressure from the National Organization for Women and other women's groups, Johnson issued a second directive in 1967. Executive Order 11375 mandated preferential action on behalf of women as well as minorities.[18] Like so many civil rights victories, women's entrance into the building trades in the late 1970s followed an African American struggle.

According to the U.S. Labor Department's Bureau of Apprenticeship and Training, in 1968 there were only two women among 77,151 apprentices across the entire construction industry.[19] In 1970, recognizing the paucity of women in construction trades jobs, the Comprehensive Employment and Training Act (CETA) program provided governmental funds to the state of Wisconsin to fund grassroots advocacy organizations to establish training programs and support mechanisms that could foster the successful integration of women in the building trades. The project created two programs for women, one that ran from 1970 to 1973 and the other from 1980 to 1983, to prepare women for entry into unionized building trades apprenticeships.[20] Unfortunately, with the onset of the Reagan era and the withdrawal of governmental funds for job-training initiatives, the Wisconsin project met a similar fate as other 1970s programs that aimed to integrate women into the building trades. The Wisconsin programs and other state-level initiatives resembled the programs that were established in the 1960s to recruit minority male workers. However, the extensive outreach and training programs created for minority men were never adequately redirected or adapted to women.[21]

One barrier was cultural. Outreach programs were designed for the underprivileged, but the women initially interested in building trades jobs were often white, middle-class, sometimes well-educated, and most committed feminists and lesbian rights activists. Another barrier was age. Women who sought building trades jobs usually had more experience in the labor market and were generally older than male applicants. The maximum age of twenty-two commonly specified for admission into apprenticeship programs disqualified many women.[22]

Women of color confronted multiple barriers. The mainstream civil rights organizations, male trade unionists, and the radical black power groups of the 1960s and 1970s all worked to advance the economic position of black men but not black women. According to Pauli Murray, an African American woman who worked on the President's Commission on the Status of Women and a member of the Committee on Civil and Political Rights, conventional wisdom held that the needs of black males were more important than the needs of black females.[23] Murray wrote: "The tragedy of black women today is that they are brainwashed by the notion that priority must be given to the assertion of black male manhood and that they must now stand back and push their men forward."[24] Black women like Ella Baker, a grass-roots stalwart of the civil rights movement, complained of sex discrimination in organizations such as the National Association for the Advancement of Colored People (NAACP) and the Southern Christian Leadership Conference (SCLC).[25] In the 1960s and 1970s, Black women labored mainly in domestic work and other low-wage service-sector jobs. They desperately needed training and access to skilled blue-collar work. But in fact—operating on the assumption that black men's needs superseded those of black women—the training and outreach originally intended for male racial minorities often discriminated against women of color.

Men in the civil rights movement were not the only barrier keeping black women from decent paying jobs. Even the director of the Women's Bureau in the U.S. Department of Labor, Elizabeth Duncan Koontz, asserted the irrelevancy in 1970 that "many women do not seek employment. Practically all males do."[26] Labor leaders and black civil rights leaders felt the same. Both viewed the inclusion of gender as a protected class with a great deal of skepticism. Civil rights leader Bayard Rustin objected to the inclusion of women in the Labor Department's Order 4, which guaranteed equal opportunity in the construction trades for men of color. Andrew Biemiller, director of the American Federation of Labor-Congress of Industrial Organizations' (AFL-CIO) Legislative Department from 1956 to 1978, wrote to Rustin supporting his objection to including agenda items from the National Organization for Women (NOW) at an impending meeting: "This must have happened when I wasn't looking," he wrote. "I do not regard them as a civil rights organization. They are interested in absolutely nothing but super feminism."[27]

Order 4 required federal contractors and subcontractors to recruit and train racial minorities with goals and timetables based on an EEOC determination of their availability in the labor force. Confronted by women's advocates such as Murray and Betty Friedan, Koontz insisted that the same standards could not be applied to women workers as to minority males.[28] "The workforce pattern of women and racial minorities differs in significant respects," she argued.[29] The Labor Department certainly had no intention of applying Order 4 to women, claiming that it "is not totally suitable to sex discrimination."[30] However, thanks to successful legal challenges like that of Lorena Weeks, a telephone operator who sued Southern Bell for refusing her application for a higher paying position as switchman, "stereotyped characterizations" that excluded women from men's jobs were deemed "romantically paternalistic."[31]

It took four more years of struggle—direct action such as picketing construction sites, aggressive advocacy, and legal challenges—to get Congress to give teeth to the civil rights law by passing the Equal Employment Opportunity Act in 1972.[32] This Act included women as a protected class, gave the EEOC the power to sue in court, and required employers with as few as fifteen employees at any work site to comply. The Act also protected public employees from discrimination and made public employers responsible for taking affirmative action to ensure equal employment opportunities for minorities and women.[33] At last the conventional notion of what defines male and female work and the exclusion of women from "men's jobs" could be openly challenged on both the national stage and the shop floor.

The Feminist Wedge

By the early 1970s, college-educated white women, influenced by the civil rights movement and the emerging feminist and lesbian rights movements, were very clear about their rights to a so-called man's job in construction, regardless of the economic plight of males of any race or class.[34] They asserted this right through women's advocacy organizations, litigation, and the press, putting unions, contractors, and governmentally subsidized job-training programs, such as community pre-apprenticeship programs, under new scrutiny. These pre-apprenticeships (in organizations such as the Workers Defense League, the Recruitment and Training program, All-Craft, and Nontraditional Employment For Women) conducted outreach and training for women and minorities and helped them to navigate the murky waters of applying for unionized apprenticeships.

Feminists saw that the liberation of working-class women depended not merely on redefining cultural concepts of masculine and feminine, but also on equal opportunities in the labor market. Female dependence on male wages is clearly a constraint on liberation, and feminists could not support the concept of a family wage reserved strictly for male breadwinners. Theories about racial segregation in the labor market might analyze the "divide and conquer" logic of capitalist enterprises, but suffer from gender blindness.[35] And they do not address the interrelationship between domestic patriarchal relations and women's continuing relegation to low-paid, low-skilled work.[36] Feminists believed that women's liberation required both rescuing a woman's personhood out of the isolation and confines of the family, as well as sexually desegregating the labor market so that women could achieve wage parity with men.

Feminists persevered. Women, especially women with children, became a growing percentage of the paid labor force. Many of these new workers took jobs traditionally held by men—from professions like law and medicine to the blue-collar construction trades.[37] These economic and cultural changes, along with a strong mandate and some financial support from the government, inspired women's groups to recruit women for the trades. Furthermore, the women entering the trades in those days were optimistic about their prospects for success.

Unionized construction craft work could provide an opportunity for a living wage. This was especially true, feminists believed, for poor ill-educated women who exemplified "the feminization of poverty," a term coined in 1978 by Wisconsin researcher Diana Pearce. She found that almost two-thirds of the poor over age sixteen were women and their overall economic status had declined from 1950 to the mid-1970s, despite the fact that more women were entering the labor force in those years. Female-headed households in particular made up a growing percentage of the poor, a trend aggravated by inadequate governmental support for divorced and single women.[38] Clearly, since the skilled trades had provided men who had limited education with good wages and benefits, with governmental backing, good training programs, and effective advocacy, getting into the construction trades should do the same for women.

Of course, women seeking to escape the "pink collar" ghetto needed counseling and preparation for these well-paid construction jobs. Women were not ready to meet the challenges of entry into apprenticeship programs. Though women had made some headway in some nontraditional manufacturing jobs, the stereotype of craft trades as men's work for which women had neither interest nor capability remained resistant to challenge. Practically speaking, it meant shoring up math readiness for otherwise well-educated women, training women to use power tools, and showing women how to deploy their physical strength to lift heavy materials. Beyond such practicalities, the issues were more complex than those faced by minority men. The male solidarity that bound unionists together and isolated women who tried to break into the charmed circle needed to be addressed. Pioneer female apprentices said that the refusal of union brothers to take them seriously was the most discouraging aspect of their training.

In response to such problems, a pre-apprenticeship program called Advocates for Women (AFW) was established in San Francisco in 1971. AFW targeted "women of all races and cultures with emphasis on low-income women who must support themselves and their families."[39] The following year, a national feminist organization called Wider Opportunities for Women (WOW) began working with women, particularly poor minority women, to secure jobs in the skilled building trades. In 1976, AFW and WOW together filed a lawsuit that sought employment goals for women in construction.[40] The lawsuit was settled out of court and led to President Jimmy Carter's issuance of affirmative action regulations for hiring women in the building trades.[41]

By 1977, the Women's Bureau was publicizing model outreach programs to support specific goals and timetables for female apprentices, especially in federally contracted construction jobs.[42] In 1978, government mandates required that women should be enrolled in apprenticeships in numbers equal to half of their proportion in the general labor force. Targets were set; for federally financed construction projects exceeding ten thousand dollars, they dictated that female participation should reach 6.9 percent by 1981—a percentage, incidentally, not yet achieved three decades later.[43]

Neither the activists nor the pioneer women apprentices themselves anticipated the antagonism their presence would provoke from union officials, contractors, and coworkers. "Breaking into construction has proved even harder than I thought it would be," declared Star Robinson, a black woman from Harlem who appeared before TV, radio, and newspaper reporters in 1981 to tell how she was fired by a male contractor from the Ferran Concrete Company at one of the largest construction sites in New York City, the $375-million Jacob Javits Convention Center. Like many women interested in the construction trades at that time, Robinson had entered her apprenticeship with great optimism and the expectation of earning a living wage. And, like many women who entered trades jobs, especially women of color, she found herself the designated "token," the minority woman hired by contractors to give the appearance of compliance with the governmental mandate of a "good faith effort" in affirmative action. Reported Robinson: "It's been almost two years since I began apprenticing as a laborer, and I'm so broke now that the telephone company is about to cut off my service. Like women in construction, I've only been able to get work on independent building sites for a few days, and then they lay you off to keep you from getting into the union."[44]

"But the contractors are only the beginning of the problem," explained a woman carpenter on the same job. "The unions, too—with a few exceptions—have resisted accepting women members, and without union membership it is very hard to get steady work."[45]

The next stage of feminist activism necessarily focused on the problems of staying on the job in a hostile environment and recruiting more women into the trades in order to reduce the isolation that made them more vulnerable to harassment on construction sites. Outreach programs sprang up across the country funded by millions of dollars made available by the Comprehensive Education and Training Act (CETA) of 1973: Women in Apprenticeship (San Francisco), Mechanical (Seattle), Women in the Trades (Chicago), Hard-Hatted Women (Pittsburgh), YWCA New Jobs for Women (Philadelphia), Better Jobs for Women (Denver), Skilled Jobs for Women (Madison), Southeast Women's Employment Coalition (Lexington), Women in Construction Projects (Boston and Washington, DC), Women and Employment (Charleston, WV), and Work Options for Women (Wichita, KS, and Raleigh, NC). Over ninety organizations from at least twenty-seven states joined together with WOW to run a Construction Compliance Task Force to aid the entry of women, especially economically disadvantaged women, into the building trades.[46]

On the national level, the task force lobbied successfully to set goals for the employment of women on federally funded infrastructural building trades' work. The Intermodal Surface Transportation Efficiency Act (1991) and the Women in Nontraditional Occupations Act (1994) provided small grants to community organizations that worked with contractors and labor unions to train women for trades work. The task force did not, however, succeed in developing an oversight function for the Labor Department's Women's Bureau so that the bureau could monitor the number of female applicants accepted and retained in union apprenticeship programs.[47]

The New York City Story

I went to college for liberal arts, left my family early, and grew up in the Bronx in a poor and working-class neighborhood, working in Chock Full O' Nuts before I apprenticed myself to a woman who was a gold- and silversmith. Then I heard about All-Craft through another woman and it sounded ideal to me. I always liked to make things when I was a kid. So I went down for the interview and failed it. They wanted minority women that weren't outspoken. Joyce Hartwell was in charge and they were looking for a certain stereotype—welfare mothers and women who had GEDs only. Then they [All-Craft] changed their minds, and Hartwell called me for the electrical union which was opening up. She told us we had to sleep out on line to pick up an [apprenticeship] application. I didn't want to do it but eventually slept out there for four nights and three days. A lot of the women [on the line] were from All-Craft. There were about 30 of us out of a line of 1,500 men.[48]

—MELINDA HERNANDEZ, ELECTRICIAN JOURNEYWOMAN

One of the fiercest battles for women over the opportunity to enter building trades jobs in the country was fought in New York City. The battle on behalf of women who expressed an interest in applying for unionized apprenticeship programs was instigated by the All-Craft program, and later by Women in Apprenticeship, which eventually became Nontraditional Employment for Women (NEW).[49]

Chartered in 1974, All-Craft got federal CETA funding during the Jimmy Carter administration. It aimed generally to provide women with an introduction to a variety of trades and to encourage their entry into construction trades. Specifically, it provided three months of workshop training in such trades as carpentry, plumbing, and electrical work. All-Craft also enticed contractors to hire women by splitting half the cost of their wages. The organization can be credited with having trained approximately 1,400 economically disadvantaged women.[50]

Using federal funds, All-Craft also established Mothers and Daughters Construction, an independent company that took on its own construction contracts. Most often it was work that no one else wanted, such as de-leading, asbestos removal, or hanging fixtures at the New York City morgue. These jobs provided women with much-needed work experience and prepared them to succeed once they gained entry into a traditional apprenticeship program.[51]

Although All-Craft began as a program primarily for white working-class women, it ultimately shifted its emphasis to minority women and men because its staff became convinced that placing more blacks and Hispanics into the construction trades would have a catalytic role in improving minority communities. The year that Melinda Hernandez slept out to become an electrician, minorities made up 75 percent of the women served by the All-Craft program. Four in five of its trainees headed households, nearly half of whom were on welfare.[52]

All-Craft soon discovered that its clients, especially those from disadvantaged economic backgrounds, faced unexpected obstacles. For example, many building trades, including Local 3, required that requests for applications be sent by registered mail—but the registered mail window at New York's main post office at 34th Street closed at 7 p.m., well before women who worked in food services or cleaning got off work.[53]

All-Craft director Joyce Hartwell described another obstacle for women: "They exceeded the traditional age-range of 18 to 22 years for apprentices—though of course when many of the interested candidates were younger than 22, Local 3 wasn't accepting women candidates at all." For example, the number of female applicants for electrician apprenticeships rose from six in 1978 to seventy-five in 1980.[54] Regardless of the growing number of applications, from 1978 to 1982 the Local admitted only fourteen female candidates. By 1982, when a significant number of female candidates were turned away from applying to the apprenticeship program due to the age limitation, All-Craft initiated a class action suit over age discrimination in apprenticeship against the Office of Federal Contract Compliance.[55] The EEOC found that the union had admitted men over the age of 35, well beyond its stated limit of 22, as part of its organizing efforts on nonunion job sites. As part of a settlement, the age limitation for apprenticeship was formally lifted in 1985.[56] And, as a direct result of the lawsuit, New York State lifted the age limitation not only for Local 3 but for all apprentice programs.

All-Craft's success in breaking the age barrier in unionized apprenticeships enabled low-wage-earning women who had more experience in the workforce, as well as poor women with dependent families, to apply for electrical apprenticeship. By 1986 the number of applications from women over twenty-five years of age comprised the vast majority of female applicants.[57]

After the age rule was removed, another barrier was erected. Education requirements for entry into the apprenticeship program were raised. Claiming that standards had to be protected, some unions, including Local 3, began requiring one or two years of high school algebra for would-be apprentices. In the electrical apprenticeship program, 74 of the 141 women who applied in 1986 were denied entrance because they lacked a regular high school diploma and other educational requirements. Sixty-six were rejected for having a low score and only one was rejected on the basis of age.[58] According to Hartwell, "These are requirements that [were] simply raised to keep minorities out."[59] All-Craft responded with another lawsuit in 1987 alleging that it was not necessary for minority women and men to have taken high school algebra if their math skills were good. The settlement of this suit also removed as discriminatory the union's requirement of a regular high school diploma rather than a GED.[60]

Despite All-Craft's successful litigation, and its militant determination to integrate women, especially women of color, into the building trades, changes in federal training program regulations during the Reagan administration soon diminished its effectiveness. CETA lapsed, replaced by the new Job Training Partnership Act (JTPA), and funding formulas changed.[61] Recalls Hartwell: "There were great difficulties meeting the requirements under performance-based contracts in JTPA. We had a high amount of dropout rates because of legitimate problems and we were dropped out from the JTPA and our applications to the unions were not accepted because they were one to two hours late.[62] Under JTPA, programs like All-Craft now only got paid for supervising a woman's training if that training was formally completed. Hartwell again: "If we took a woman from the shelters that had children to support, and her children got chickenpox and she had to drop out of class, we had to pay back the money that we received to train her. When we had a high dropout rate, we could not pay our bills." All-Craft lost the training contract for its Mothers and Daughters Construction Company in 1986.[63] The agency turned to sheltering and treating homeless addicts.

Noble work to be sure, but a sad turn for an organization that had done such excellent work teaching women the construction crafts.

All-Craft's shift in focus certainly affected the density of women seeking electrical apprenticeships. By 1987 the number of female applications applying to the apprenticeship program dropped back down to seventy-five. By 1989, the Joint Apprentice Committee (JAC) took applications only from women likely to be accepted. Six women applied and three were accepted. Sandra Pardes, a young Puerto Rican, was one of them: "I applied to the apprenticeship program of the IBEW Local 3 in the late spring of 1989. Of the approximately eighty people who showed up, I was the only woman."[64]

Reliable and consistent data on minority and female application and acceptance rates to building trades apprenticeships, including electrical apprenticeships, became impossible to obtain after the Reagan administration shifted reporting oversight from the EEOC to JACs in each of the building trades. These JACs are obligated to report only the number of minority and female apprentices already enrolled in their programs to the New York State Apprenticeship Council, the agency that has oversight responsibility for registered apprentice programs.[65] But data published by the New York State Department of Labor and other sources show that women have not made any significant gains in the brotherhood's electrical apprenticeship in New York City.[66] Although the state—under pressure from women's advocacy organizations—set a goal of 45 percent female participation in the building trades during the 1980s, the actual enrollment of Local 3 female apprentices in New York City during the decade was an average of 2.7 percent—1 percent higher than the ironworkers but slightly less than the unionized carpenters, plumbers, and steamfitters.[67]

In 1992, women comprised only 0.07 percent of the journey-level union membership of Local 3. Over the rest of that decade, the overall percentage of female electricians, including journey-level and apprentices, ranged from 1.1 percent in 1993 up to 2.2 percent by 1996, and then back down to 1.9 percent by 1999.[68] Reportedly, as of 2007, the JIB and Local 3 claimed that there are 300 women—journeywomen and apprentices—out of 13,000 Division A construction electricians.[69] Women in New York may be 52 percent of the population, but they remain a mere 2.3 percent of electrical workers.[70]

After All-Craft's demise, another feminist entity, Women in Apprenticeship (WAP), joined the quest to help would-be journeywomen. Initially funded by federal CETA dollars, WAP started with a less confrontational agenda, aimed simply at informing women when unions were accepting apprentices and where they could get applications. In line with union rules, WAP turned away women without traditional high school diplomas. When CETA ended, WAP became NEW, and sought alternative funding.

Strategically, it began to stress cooperation with the unions over its previous antagonistic stance. Currently, NEW has a strong cooperative relationship with trade unions based on its willingness to eschew high-pressure tactics and to integrate women's issues into the broader frame of trade union issues.[71]

This cooperative relationship both facilitates and hampers the organization's ability to integrate women into the New York City building trades. NEW does a fine job recruiting and training low-income and poor women, much like the anti-poverty programs promulgated by the feminist movement of the 1970s which aimed to integrate minority women into the building trades. On the other hand, NEW relies on the New York City building trades unions and the New York State Building Trades Council—the very organizations that historically have resisted the entrance of women as a critical mass in the industry—for political clout and funding, and this greatly hampers NEW's ability to call in governmental agencies or litigate to enforce compliance.

Programs like NEW certainly do provide job training to poor and low-income women, and the successes and importance of innovative training opportunities for noncollege women who end up in low-paying, low-mobility, pink-collar jobs are well documented. Researcher Sharon Mastracci examined the relative effectiveness of various programs in helping women (like those in NEW) gain access to high-wage, high-mobility employment opportunities. Using case studies of grant-funded projects and in-depth statistical analysis of ten years' data on women throughout the United States, she concluded that intermediary organizations are critical to transitioning low- or no-income workers into higher paying, more secure jobs.[72]

NEW conducts eight-week pre-apprenticeship programs in the trades, including carpentry, plumbing, electrical work, masonry, pipefitting, and steam fitting. NEW's graduates are over 60 percent African American, 20 percent Hispanic, and only 10 percent ethnic white. Seventy-four percent are single heads of households and 60 percent have worked for several years in such low-wage service-sector jobs as waitressing, administrative support, and hospitality. Fifty-five percent were on some form of public assistance. Therefore, the typical graduate of the NEW program is an African American female single head of household in her thirties with school-age children to support. Unfortunately, ninety-day retention rates appear low.[73]

NEW is now the only remaining program in New York with the mission of integrating women into the building trades. As such, it faces formidable challenges: limited funding due to federal cuts to women's job training programs, political boundaries and legal constraints imposed by the contractors and unions, and dependence on male directors of JACs within each trade to place women graduates into apprenticeships and jobs.[74]

The Local 3 Story: Women Meet the Culture of Brotherhood

When I went for my interview [to the JIB] I noticed some older white men with their arms wrapped around younger white men—they appeared to be fathers with their sons. The fathers would talk to the men at the sign-in table and then talk to the ushers who brought you to the interview table. After talking with the ushers, these young men were immediately taken into the interview room. They did not follow the same waiting process that I and other applicants did.[75]

—SANDRA PARDES, JOURNEYWOMAN, LOCAL 3

In the historically bitter relationship between labor and capital, both sides have often found oases of common ground in the stereotyping of women's roles. Scholars such as Ruth Milkman in her study of the auto industry, Cynthia Cockburn in her work on male printers in England, Patricia Cooper in her study of unionized cigar makers of the early twentieth century, and Ileen DeVault in her recent work on gender polarization in late nineteenth- and early twentieth-century unions have demonstrated just how complex and contingent is the history of the sex-typing of jobs as "male" and "female."[76] An industry's formative period is critical in the process: a pattern of employment by sex, once established, soon becomes extraordinarily inflexible. Employers prove unlikely to challenge it, even at the cost of lower profits; as noted in Chapter 1, ordinary men and women come to accept the divisions as, simply, natural. The electrical industry is no exception.[77] Certainly the electrical brotherhood has managed to severely restrict female access to work as much as or more than any other New York building trades union.[78] Members and leaders of Local 3 may take comfort in "this is how things have always been done," or suggest that "if it ain't broke, don't fix it," but few understand how this fear of change—this inflexibility in a contingent economic world—may ultimately undermine their interests. Even in the twenty-first century, Local 3 clings to the ideal of the male breadwinner as the sole provider of his family, and uses this ideology of brotherhood as a shield against heightened competition, rapid technological change, and expanding use of low-wage nonunion labor in the construction industry. The family-wage ideology of the early 1900s, which assumed female dependency, was reinforced by the gender conservatism of the 1950s and 1960s, and is being reinforced more recently by a backlash against feminism.[79]

The attachment to this male-oriented ideology was especially intense among electrical workers after World War II, when the increasing numbers of women entering the labor market were seen as subverting the argument for a family wage and threatening the very idea of a union.[80] As Kessler-Harris pointed out, blue-collar wages high enough for a single male breadwinner to support a family—wage-gains that are the fruit of a long and noble struggle—paradoxically make it unlikely that the value of women's work outside the home, measured in their wages, will reach equity with the work of men. If that happened, it would shake the psychic foundations of their ideology of manliness itself. The ideal of the family wage persists to this day among tradesmen, even among young male apprentices. They accept that their wives or girlfriends may work outside the home, but not the prospect of their women learning a skilled trade equivalent to their own. In fact, they prefer that their wives take responsibility for the children and the house while they dedicate much of their leisure time to voluntary union activities. To this day, the Local honors the wives of electricians from the Division A construction group for their supportive role at home at an Annual Ladies Night, a dinner where mink coats and other "feminine" items are raffled off.[81] The Local hosts this event in order to encourage the wives to take on more domestic work at home, freeing their husbands to perform volunteer work for the union.

A male journeyman's views of women's roles in and outside the workplace are becoming increasingly complex, but this complexity is still infused with gender and racial assumptions. Wives of European American electricians are, for the most part, homemakers and nurturers. By contrast, the brothers can readily accept African American women in electrical work as lower wage industrial workers and "sisters" in the brotherhood.[82] Although generally confined to the manufacturing divisions (e.g., lamps and lampshades), these women are even occasionally promoted to active positions in the Local.[83] Further, although African American women do experience racial harassment on construction job sites, white male electricians rationalize their presence as the failure of African American men either to hold good jobs or to marry their women.[84] Although men in Division A generally resist the presence of any women on construction job sites, the union leadership is proud of the fact that sons of minority women from the manufacturing division are recruited into the Division A apprenticeship program. This mother to son recruitment strategy not only leaves undisturbed the racial and sexual hierarchy of the brotherhood and its paternalism, it serves two useful purposes. First, minority electricians are made available to unionized contractors for federally funded projects. Second, minority male apprentices "mothered" into the Local seem more easily disciplined. Since women in the manufacturing division rely on the Local to fight injustices on the shop floor, they are more likely to forewarn their sons about the economic and political perils of causing problems for the Local with contractors on job sites.

In its dominant ideological assumptions, the brotherhood applies a double standard to women. Typical of the racially exclusionary character of early twentieth-century fraternal associations discussed in Clawson's work, the dominant gender expectations of Local 3 are race-driven. White women are expected to maintain an unambiguous gender identity as wives and nonworkers, while black and minority women in subordinate divisions can attain status, albeit second-class status, as workers, union members, and mothers of future minority journeymen.[85] Only the rare minority male trainees who enter the apprenticeship program through a route other than the Division A apprenticeship program rank almost as low as female workers.[86]

Privilege in job assignments, overtime work, promotions, and decision making is given not just to white males, but specifically to male descendants of the original Loyal 100, as well as to the sons and nephews of Division A journeymen. Male culture and solidarity among union electricians is pervasive. It is embodied in the Local's gendered divisions; fraternal social clubs; labor and management board; JACs; and the initiation, training, and educational components of its apprentice program. Each of these aspects of organizational life of the Local and the JIB are attempts to hold onto the union's power vis-à-vis employers and define and reinforce the meaning of brotherhood—even as the definitions of gender identity and social roles remain in flux in the larger society, and even as Local 3 has had, to some degree, to bow to antidiscrimination laws.

The First Class

Local 3 accepted its first woman electrical apprentice more or less by accident. Melissa, an African American from the Recruitment and Training Program in Harlem, was admitted to the electrical apprenticeship as a minority male in 1974; then, after a sex change operation, she was readmitted as a minority female in 1976. She became one of America's first female top-level mechanics in electrical construction, and a celebrated and rather "notorious" member of Local 3's Division A. But the first actual class of women applicants applied in 1978.[87]

On that day, June 26, thirty women waited in line at the JIB's office on Jewel Avenue in Flushing, Queens, hoping to be accepted in Local 3's 1978 apprenticeship program. The press was there to report how Title VII's provision on sex discrimination played out on the streets of New York. At least one would-be apprentice, a Puerto Rican woman, Melinda Hernandez, had slept on the line for four nights to preserve her place. Anonymous men menaced the women; one urinated on a woman's sleeping bag. But the women held firm. Only six were accepted; two soon bowed out. The remaining four—Asian, African American, Hispanic, and Jewish—became a hard-fighting and supportive cadre within a very hostile environment.

Their acceptance into the apprenticeship program had been preceded by struggle: feminist organizations had filed lawsuits, politicians and bureaucrats had changed policies, and all the while construction contractors and union leaders balked. The four pioneers met resistance every step of the way. Because these women had been referred to the program through the efforts of feminist agencies, they did not have even the minimal social mechanism of a father-to-daughter legacy to protect them against ill treatment. Their efforts to develop positive work and social relationships with male coworkers were often construed as sexual advances.

These four are the foremothers of all the women who followed in the electrical trade and in other construction trades and they personify the heroic struggle that tradeswomen waged against whiteness and maleness in craft occupations in the 1970s and 1980s. Each of them had been referred to apprenticeship by All-Craft. The second year's group was even smaller: two white feminists, Laura Kelber and Evan Ruderman. One of them entered by a variant of father-son sponsorship; Laura was the daughter of the director of labor education at the Local. When she first became pregnant and informed her employers, she was assigned heavy lifting duty and suffered a miscarriage.[88] She finally left electrical work in 1998 after filing a lawsuit against the contractors for retaliatory action. Evan worked with her tools up until her death from HIV/AIDS in 2003.

The one Jewish woman in the first class—Beth Schulman—eventually left the industry.[89] Cynthia Long, Melinda Hernandez, and Jackie Simmons stayed with their trade for over twenty-five years, struggling with their employers, the Local, and the conventional wisdom that women just did not belong in the industry. Cynthia is the only remaining pioneer working with her tools out of the original four women. All four pioneers, however, helped mentor and teach subsequent waves of women. Men accepted the women apprentices as tokens—like their counterparts the carpenters, plumbers, and operating engineers—only when it became obvious that the entrance of women was inevitable because the women had federal law on their side.[90] That did not mean, however, that men would welcome women into the brotherhood or teach them the trade.

One journeywoman electrician who entered the trade in the early 1980s recalls:

> As soon as I entered the apprenticeship program, I found much to discourage women from continuing. Women weren't welcome. No toilet or changing facilities were available. Some men treated women as invisible. Those assigned to teach women did not go into detail. We had to force ourselves on them to be able to learn. The males did not have to do that. They were treated with respect.[91]

It was an inauspicious time for the women apprentices. The recession of the mid-1970s had hit the construction trades especially hard. By 1978, when the Local was beginning to emerge from the bad times, many male craftsmen were summarily called back from jobs around the country where they had settled with their families, and were forced to relocate to New York City on pain of losing their union cards. The fact that this occurred just as women were entering the industry, as were more African American, Hispanic, and

Asian men, made the new entrants, especially women, scapegoats for collective anger and bitterness.[92]

As noted above, women who initially entered the electrical trade and other building trades belonged usually to one of two groups: poor and working-class minority women, and white middle- and upper-class college educated women—sometimes Ivy Leaguers—who were feminists or even lesbian rights activists. Mary Ellen Boyd, co-founder of WAP and former director of NEW, observed:

> The first women who went into the building trades, especially electrical work because they thought it was "brainier" than some of the other trades, were truly upper class and white. They were women who mostly went to college and saw the work as an extension of their feminism or their freedom to function in a different way. They genuinely wanted to do physical work or craft work. That was in the late 1970s and the first group of women since World War II.[93]

Cynthia Long estimates that more than half the pioneers entering construction were lesbians, which also caused rifts later on among tradeswomen.[94] In the 1980s, after All-Craft's successful age discrimination litigation against the union, "older" women complicated this demographic. The minority women who were poor or working class had toiled at minimum wage jobs as secretaries, daycare workers, and waitresses, and they wanted the job security, wages, and benefits afforded workers who entered unionized apprenticeship programs. College-educated women most often sought electrical work as a countercultural expression. Class and racial tensions ensued. Boyd was actually asked by some white college-educated tradeswomen to stop bringing poor women into the NEW program "because it was hurting their image."[95] Boyd recalls: "They thought poor women didn't have an equivalent speaking capability—they sounded rougher or whatever. And poor women wouldn't handle relationships between women and men on the job the same way they did."[96]

Whatever their differences, women pioneers and those who followed through the late 1980s were united in their experience of profound isolation, stress, and workplace harassment, both sexual and nonsexual. The fraternity of union electricians simply would not accept women as colleagues. A woman-of-color electrician journeywoman recalls: "I began working in construction in the early 1980s. I heard about a CETA-funded project which helped women get into the unions. I waited on line for several days. . . . Male applicants told us that women didn't belong in this business, and that we were taking jobs away from men who had families to support."[97]

In 1989, ten years after entering the electrical industry, at a reunion of women in the trades in New York City, Evan Ruderman recalled:

I worked six months as a nonunion apprentice and then Local 3 came around and threw me off the job. The first days when I came into Division A of the electrical apprenticeship program the biggest problem was bathrooms. My family would complain, "oh, Evan is going to tell another bathroom story." After ten and a half years, it hasn't changed enough for me. Currently, conditions in construction are very bad and the excuse is that the contractor has to stay competitive. When it comes to fringe benefits for the Local members, the leadership never complains but if you raise an issue of advocacy for women on the job sites and for women's rights in construction, then the union leadership will tell you that they have to keep the contractor competitive. I have found that people will rise to their expectations and currently the expectations of men in construction are not very high.[98]

Women's attempts to have their grievances addressed by the union were, for the most part, futile. A journeywoman electrician testified anonymously at the 1990 New York City Human Rights Commission Hearings:

Attempting to take problems [on the job site] to the union leadership doesn't help. A group of women asked to meet with the apprentice director to address the problems of sexual harassment and lack of toilet and changing facilities. He told us we should just deal with it. A proposal by women apprentices that union reps, stewards, and others in management take steps to correct the problems was ignored.[99]

After the Clarence Thomas hearings in 1992, however, Local 3 recognized that sexual harassment could create personal liability and scandal for men at all union levels.[100] Therefore, during the 1990s, the union made an adjustment to its "all in the family" model of recruitment into the brotherhood. To the dyads of "father to son" and "mother to son" were added "father to daughter," "uncle to niece," and "brother to sister" as ways to recruit "safe" (nonfeminist) females into the Local's Division A. Even if they were related to the male members, women apprentices were always taken on in very small numbers—never exceeding more than about 2 percent.

Women who entered Local 3 via the family route (both real and fictive) had it somewhat easier than the pioneers. They entered the trades less as a feminist project and more because they had a relative in the industry or simply needed to support their families. In the new century, women admitted to the Local mostly enter through a connection to a relative in the industry. Though closely connected to union men, these women still experience the usual isolation, sexual harassment, and discriminatory behavior on the job. However, unlike their pioneer sisters, these "connected" women are somewhat alienated from the pioneers and less likely to openly challenge contractors, unionists, or male coworkers. As Stan Brown, an African American journeyman, observed:

> The type of women who come into Local 3, I would say that 35 to 45 percent are gay, or women from families of electricians (they usually don't have trouble in the union) because psychologically they are better prepared, they tend to identify with the organization. The irony is that these women are in the strongest position to raise feminist issues in the Local and yet they reject gender politics.[101]

Of course, pressure from any outside group on the brotherhood has always been exceedingly distasteful to the union. But, as Harry Van Arsdale explained, "Instead of fighting change, it's better to get involved, and in that way you can exert some control."[102] However, as Evan Ruderman angrily asserted:

> The leadership owns the union. And therefore, it is their club. They decide who comes into that club, not some outside force such as the state or the feminist movement. All this is in order to control the members. Once you are admitted to the club, the male leadership decides what you do and how you are going to participate.[103]

The Local may declare that anyone can join the "club," but this is a dodge. Few women are sponsored, and while those who are connected may have a better chance at sponsorship, they never achieve any leadership positions.

To the members, this is "brotherhood" in action. Construction electricians pride themselves on the physical nature of their work. Their very identity is bound up in their doing, as one puts it, "the work that only men can do."[104] Building something implies power and skill, attributes still generally construed as masculine. Union men want to keep it this way.

Of course, brotherhood is defined by more than mere physical strength. It is also a code for the traditional expectations of stability and familiarity. One typically conservative white male electrician told me: "Unlike blacks or Latin men coming into the union, women break up the existing pattern. Here is new power coming onto the scene. They are affecting order on the job." He points out that men are already linked into networks before they become apprentices: "Either you are an older 'Roger Bacon' boy from an all-boys' Catholic high school, or you might be a former 'jarhead,' that is, you were in the Marines." By way of contract, he adds, "When women come in this adds a new dimension—they don't have knowledge of the industry, they don't have the understanding of the way men relate, but they come in with a lot of power because all they have to do is 'wink their eye.'"[105] As another male journeyman pointed out: "[The men] have no respect for [women's] 'workability' or for them as tradesmen. They just don't see women—they don't see a place for them in the industry, so they'll just ice them—they'll block her out—make her feel uncomfortable and let her know that she's not part of the group."[106]

Many journeymen express a fear that once women are admitted freely, they will take over. The fear is obviously exaggerated. In thirty years, women have made no significant gains in electrical work or in any of the construction trades.[107]

The Trouble with Travelers

An irony often noted by observers and analysts of discrimination is that the worst abuse a despised minority suffers comes not from those on the top or in the middle of the status hierarchy, but from the group just above the bottom rung. Beaten down by everyone else, this group cherishes its prerogative to beat down the one group below them, even though their political and economic interests may coincide. This was the dynamic between poor whites and African Americans in the Jim Crow South. On the electrical construction job site, relations play out in just the same manner between women, on the bottom rung, and the "travelers" just above.

"Traveler" designates workers who travel to work sites around the country. They are the itinerants of the electrical construction trade. Travelers, unlike Local 3 journeymen, return to their home districts after a layoff. They are an important component in the union's ability to maintain its jurisdiction in the industry. During a slowdown in construction, more union members become unemployed. Prolonged unemployment can lead Division A mechanics to take nonunion work with corporations such as AT&T and IBM, often in a practice called "double breasting" where both union and nonunion workers do the same job.[108] This is not only an unfair labor practice, but contractors and unions train workers at some expense, only to lose them to companies which invested nothing in that training.[109] Because travelers can be sent home when jobs are scarce, leaving the available work for local journeyman, it gives unions vital flexibility in regulating their labor supply and reduces the risk of their members going to work for nonunion employers.

Nevertheless, in Local 3, travelers are second-class citizens. They are assigned the most physically taxing and dangerous work—in the subways, at pollution control plants, and at large-scale outside deck jobs where they drill off the sides of buildings, chop concrete overhead, and remove asbestos. They are at the bottom of the list for overtime and rarely become foreman or shop steward. They cannot request rotation to another job site. They do not share in the local union's fringe benefits and are not full members of the local. They are less likely to gain a coveted journeyman's card, although they pay the same union dues as their local colleagues—and they may even pay a premium to host unions to work in certain areas.[110] Despite these disadvantages, for male workers who are unable to support their families in their home districts, traveling is an alternative that allows for economic survival. This traveling tradition is as old as the trade itself. But travelers are strongly resented by Local 3 electricians. As one Puerto Rican male Local 3 electrician stated, "We have enough unemployment, and we have enough young men and women who we could put to work from here right now. You know we could change the lives of an entire family with a job, and we keep hiring these out-of-town guys."[111]

Travelers come mainly from the nonunion South and Southwest, and perhaps this inspires even more resentment. Their outsider position in the union should make them somewhat sympathetic to minority male workers and women, but their racist attitudes are well entrenched and foster a climate of mistrust and hostility. Travelers are strongly resented by Local 3 electricians. One female African American fourth-year apprentice found that it could be more difficult to work with travelers than with her local white colleagues:

> I had a hard time with white out-of-towners. They seem to give women and black apprentices more problems—not the journeymen, just the apprentices. There was a white man from Mississippi and he told my black partner that he would not work with him. The white foreman ordered the traveler to work with my black partner—and the traveler left the job site. He would rather lose a day's pay than work with a black guy.[112]

The travelers have a dissident traveling fraternity of construction electricians which reportedly is as old as the IBEW itself, and called the Federation of Lineman and Electricians (actually FLE but referred to as FLEAS). Their logo shows a small flea holding a traveler's knapsack. FLEAS are highly critical of the established union structure. Contractors and union officials fear that FLEAS may be too militant to be easily controlled. As one traveling journeyman stated, "The FLEAS are the guys who get up and talk at union meetings. These are the guys who are the critics of the union."[113]

Many women are critics too. But any alliance is unlikely. At the annual FLEAS picnic, I spoke with members who claimed to identify with the Hell's Angels and the Ku Klux Klan. They are more likely to scapegoat women than join with them in solidarity. Women, especially women of color, appear to have no natural allies within Local 3—not even on its margins.

4

On the Electrical Construction
Work Site: The Sexual Charge

At the work site itself—the great clanking, churning, cacophonous hive of steel and concrete and wire—the economic imperative to build meets the social system of the brotherhood. Yet the focus here is on the determined and checkmated women electricians of the 1990s and today who struggle to gain a foothold in Local 3's Division A: white, black, and brown; lesbian and heterosexual; middle-class, working-class, and poor; the generation that followed the initial cadres from the late 1970s and 1980s. Some three hundred women electricians entered the Local 3 electrical program in these decades, out of thirteen thousand union apprentices.[1]

These women are not uniform in their attitudes and approaches to the union. Approximately half of them entered the industry and the union under the auspices of feminist and minority-based community organizations dedicated to recruiting a new workforce into construction. However, even as All-Craft was taking Local 3 to court in the 1980s, another channel for women opened up.[2] Through the father–daughter connection, women could be referred to the apprenticeship program by male relatives or boyfriends already in the union.[3] Many women who came into the union this way believe that feminist activism outside the arena sanctioned by the union may rebound to harm them.[4] While one group owes loyalty to outside groups that the union considers reluctant partners at best and adversaries at worst, the other owes loyalty to the traditional union organization.

So there is tension and division on the work site even among women. While construction craft unionism is constructed on the ideological base of brotherhood, there exists no alternate model of sisterhood to support women workers and challenge traditional practices. Gender solidarity is also inhibited by the simple fact that there are not enough women on any particular job site to create a critical mass.

As noted in Chapter 3, among the pioneers in the late 1970s and 1980s, a sizable plurality of women in construction came from educated, middle-class backgrounds and were motivated to join the trade for ideological reasons, either countercultural rebellion or feminism.[5] Later, the majority of women came from blue-collar families. If their mothers worked, they held traditionally female jobs like clerk, nurse, and seamstress. They themselves mostly worked as waitresses, clerical workers, fast food servers, secretaries, daycare assistants, and receptionists.[6] Such positions pay much less than construction work; some pay only minimum wage. These women were more interested supporting themselves and their families than in ideology.[7]

Across the entire economy, women comprise almost half the workforce, but they are rare in the skilled trades; only 2 percent of the 5 million workers in skilled trades are female, a common pattern in well-paid nonmanagerial occupations.[8] Men resent the women entering the high-paying blue-collar positions because they lose a monopoly on two precious commodities—their high-wage jobs and their gender privilege at home.[9] Women from blue-collar families (especially with family members in the trade) tend to shy away from confronting this resentment. At the same time, they distrust the feminists, who seem too often to seek confrontation and publicity.[10] The tension between the two groups is unfortunate as their interests are, after all, identical—ending the feminization of poverty.[11] As of 2008, 80 percent of American women worked outside the home; half of all mothers work. Seventy-five percent of divorced mothers work.[12] Nontraditional jobs provide women the opportunity to earn a man's wage.[13] After nearly two decades of feminist advocacy for poor and working women seeking entry into nontraditional work, the average woman still earns only 79 cents for every dollar earned by a man.[14]

An Electrical Construction Workday

Work on big construction projects is like factory work: the individual worker has little control over the process. One tradeswoman describes it: "You'll walk into a room, and there will be six pipes stubbed in there and the boss will come in and say, 'Take pipe one from point A to point B. Take pipe two from point B to C, pipe three from C to D,' and so on. You say, 'Where are the pipes going? What are they for?' And he'll answer, 'You don't need to know. Do what I told you.' You could do this for years and have no idea what you have done." This process of working as part of the collective with no control over the work process is mirrored in the union member's relationship with the Local, which frequently calls on members for help and support, often without initially providing a specific reason for the call (e.g., when staking out nonunion work in one of the city's boroughs).[15] However, because it is for the union and not for some boss, Jane, a master electrician and currently an electrical contractor, doesn't mind this sort of call. She says proudly, "You feel like a piece in a puzzle, and when the puzzle is all finished—then you'll see your contribution."

The work is not easy. Male electricians talk about how much strength is required, but female apprentices say they can do the work, especially with the newer materials and tools, which tend to be lighter than the older ones. If electrical construction no longer requires brute strength, it is still uncomfortable, demanding, and dangerous.[16]

The most dangerous jobs are the "deck jobs": large, high-rise construction projects in which most of the work is conducted in the open air.[17] Deck jobs require careful training and preparation. According to the "logic" that women can't handle the hardest jobs, foremen should never assign women to deck jobs. Yet when women arrive on the job site, that is what they are usually assigned, in part to help contractors fulfill federally funded job goals, and as if to test their ability and endurance.[18] Conditions on deck jobs can be brutal, Evan Ruderman stated, "You're out in the middle of nowhere. When I say nowhere, I mean there may not be a roof, there may be no walls. There may not even be a floor. You can be climbing steel up the side of the building and it literally is nowhere because it's not on any map yet. It doesn't exist until you're finished with the job."[19]

Jane, who has worked as a traveler in Local 3 for more than ten years, continues to describe a day on a deck job:

> I get up between 4:30 and 5:00 AM and try to be out of the house by 5:30 because I need to be on the job between 6:30 and 7:00. It's real nice for beating traffic but real cold at that time of morning. You have to dress in layers. You start with your skivvies, your underwear. You'll have long 'handles' they call them, on the pants with the little drop seat in the back. Then you go to some type of light T-shirt. Then you put on a thermal shirt and a pair of jeans. If the weather says it's going to drop below freezing, then you go to the next layer, a flannel shirt. Then you'll have a sweatshirt, a pair of Carhartts (Carhartt bibs are similar to overalls for work wear), and a coverall. If the weather report says the wind chill is going to be below zero, then you add one more layer on top of that. That's when you bring out the heavy coat. You put on what we call "Arctic wear." By now, you have so many clothes on, you look like a teddy bear because your arms won't come down to your sides. They're literally stuck out at the sides.

Because construction work is dirty, it is essential to store work clothes on site. Women also need space for personal items like handbags and backpacks. But space is often inadequate and awkward to secure. "If you're fortunate," one tradeswoman says, "men have their own locker and the women have their little locker." "If you're unfortunate," she continues, "and have your 'monthly blessing,' you could be in for a day of humiliation."[20] On deck jobs usually there are only chemical toilets, and minimal locker facilities.

After stowing their personal items, Jane continues, describing her day, "we all fall out and appear in front of our supervisor's locker and the workday begins." There they receive their "brass," small numbered metal tags that the supervisor uses to keep track of the crew. This rather primitive system also helps with the bookkeeping: as the worker gets handed her "Brass 62" or "Brass 13," the number is checked off on a ledger which determines who gets paid for the day. And while it is possible for men to bargain with each other to trade work assignments, such exchanging is nearly impossible for women. The supervisor keeps their brass in a separate pile, and since the pile is so small, it is easy for the boss to make sure that each woman takes the job to which she has been assigned. The practice also keeps the women conspicuously apart from the brotherhood.

After job assignments, a worker joins a crew of two to nine others, supervised by a straw boss. The individual worker will be "tooled up" with a partner; an apprentice is usually paired with a journeyman. The setup is a practical one. Veronica Rose, a former Local 3 apprentice who is now a master mechanic and owns her own contracting firm, observes that 20 percent of the electrical industry is becoming obsolete every year, "so when you have two people, the contractors feel like there's twice as good a chance you're going to be able to install whatever they give you. A lot of times you'll see an old timer and a real young kid. The idea is that what the old guy hasn't seen, the new guy has already learned in school."[21]

Once the jobs have been assigned, then the work begins. If the task is large, such as running ground wire up the length of a tall building, the crew might join nine electricians in a whole gang supervised by one or two foremen. For every three or four gangs there will be a general foreman, possibly with a few assistants. A ground wire is a lightning rod, taking electricity that hits the top of a skyscraper down the side of a building and dispersing it safely into the ground below the basement. Running it up the building is an especially difficult task whose preliminaries involve getting 3,000-foot spools of copper wire into position—hauling them by hand over dirt, "just like building the pyramids," as one journeyman described it.[22] Then, in order to spin off the wire, the electrician inserts a big piece of pipe called a "pencil" through the metal reel to allow the wire to be sent up the height of the building with strategically placed jacks. When the wire is stretched out, the end is wrapped around a steel rod dug into the ground underneath what will become the basement of the building. After the end is buried, the wire is secured every six to ten feet—an awkward task Jane describes as "sewing backwards." "You're putting a stitch around the wire to the steel." Once it is secured, the ground wire is exposed for several months before it is covered with concrete. During this period, it must not be broken, and it must not be touched during stormy weather.

Ironworkers come right behind the electricians, installing the "skin" of the building onto the outer beams. Once the skin is installed, the ironworkers and electricians, along with the other tradesmen, perform a "topping out" ritual in which workers place a small tree and an American flag at the top of the skeletal building. Every tradesman on the construction site comes to the top of the building for this ceremony, and their pride is palpable. The owner may join the contractors and subcontractors to congratulate the workers for completing this critical stage. Jane chokes up describing it:

> When you get done, you can't believe you did it. For a woman, it's like having a baby. It really is, because you created something. . . . You have a child and she's raw and she's unadulterated. She's so innocent. Then you give her all the things she needs to be an adult. A building is the same way. It starts out really, really pure and my job is to give it everything it needs. I have to make sure it has all the power and all the lighting that it needs to function as a productive member of our society, as a full grown building.

After the topping out ritual, electricians install the "core"— the proverbial nervous system of the building—according to the special power needs specified either by the contractor or each tenant. Electricians also install the panels for the heating and air conditioning systems. There can be profound satisfaction in all this, as well; Jane says that once every corner of the new building has light and power, "you feel like you died and went to heaven." This is, of course, partly because now she and the other construction workers can enjoy heat, inside running water, and toilets. The satisfaction is in the result, not in the work.

Jane maintains that anyone could accomplish this phase of the job without a day of apprenticeship training. "The technology has advanced to such a point that if you can read directions and you have self-confidence, there's very little prerequisite. Eight out of ten panels I install, it's the first time I've seen it in my life, because they're so different. But they all come with instructions." The detail work that completes the job requires more expertise, but on the whole, this is grunt work. This dumbing-down process is known among scholars as "de-skilling." Most industries experienced de-skilling in the twentieth century, but it only recently began in construction work.[23] De-skilling is the bane of the craft worker's existence, threatening his status as a skilled worker, and ultimately the union's power to control the labor supply.[24]

The straw boss on a crew assigns workers the same jobs, day after day. They find themselves doing the same tasks on successive floors for a year or more. Monica says, "As soon as you know what you're doing, then that's what you're going to do for the next 40 or 50 floors." This allows for consistency in the work tasks, the tools, the materials, and the training. Although the process is monotonous, production is efficient. Yet it is difficult to derive a sense of accomplishment from this sort of repetitive work. It is work, the stereotypical male construction worker might mutter under his breath, that "even women could do."

The Gender Politics of Danger

Other building tradesmen consider construction electricians to be elite because of the danger they face on the job. Despite the local's emphasis on safety and training, at least one in five electricians is injured over the course of his or her career.[25] Members of Local 3's special division of workers trained in skyscraper construction receive what is called "high-time money" for working 400 to 500 feet in the air at tasks that the union considers to be extremely dangerous. High-time money is time and a half. One high-time task is installing lightning arresters around the roof.[26] This is done well before there is any completed building structure. High-time money is also available for work on bridges, replacing bulbs, servicing lighting, power equipment on heavy construction equipment such as cranes, installing lighting in high ceilings, and electrical work inside elevators.

The additional money is well deserved: the closest thing to some of this work is rock climbing, which people do for recreation precisely because it feels dangerous. Jane describes working on an elevator shaft:

> I had to go up what was a riser, and I would open the door, and there's a room but there's no floor and there's no ceiling. It goes down 30 floors and it goes up 30 floors. There's nothing in there. I go in, I drill a couple of holes, and I put a hanger up, which is just a piece of metal. "Okay that's done." Then from where I am, I put my second hanger up. When I put the hanger up, I secure a strap to it. The reason I put a strap on it is so I have some place to tie my rope. I create a ladder myself that I climb up, and as I climb it, I secure it to the pipe. It has to hold my weight—or I fall thirty stories.

It is not only the obviously dangerous jobs that lead to accidents; routine work can be just as deadly. Jane recalls an accident in which a journeyman was in a closet with other electricians testing equipment. A switch that had not been properly manufactured or installed surged to 1380 volts, more than ten times normal household voltage. The journeyman was touching the switch at the time. Veronica recalls: "He walked out of the closet with his fingertips smoking. His feet were smoking; smoke was coming out of the collar around his shirt. His eyes were wide open. He looked like a zombie. He walked out and fell on his face and he was dead. There was no doing anything."

In dangerous and unpredictable situations, even the most conscientious workers can be vulnerable, but most male workers customarily attribute injuries and accidents to individual carelessness or lack of judgment. They almost never complain about the inherent dangers of the job. In an unstable construction economy, with union workers worried (and with good reason) about being replaced by cheaper nonunion labor, to raise concerns about unsafe work conditions can threaten job security.

Danger in electrical construction is also infused with the politics of gender. Men cope with the risks by reverting to machismo.[27] A worker named Michael claims that some guys even walk the steel when they don't have to. "They like the image and they like the risk."[28] For men like this, the best way to deal with risk is not to confront greedy or indifferent employers, but to "be a man" about it, masking one's fears and taking the high-time money. Michael says, "The contractor says, 'I can't give you a safe job but I will give you an extra dollar.'" Michael seems proud that men will take big risks for big money, while most women workers will not. Risk-taking is part of the macho brotherhood ethos.[29]

Similarly, union men respond with machismo to the slights of class snobbery. They shrug off the stereotype of construction workers as "dumb." However, regardless of the individual cultural attainments of any construction worker, to say nothing of his actual economic position, middle-class social status remains unavailable to him. Former journeywoman Evan Ruderman observes:

> The guys are working and earning big bucks, but they are still not considered fit to associate with people in suits. That's told to them over and over in all kinds of ways. They even have to use the service entrance when they enter a building—even when they arrive in the morning in clean [nonwork] clothes. The message they get is that they are not fit to mingle with minimum-wage secretaries in the elevator. They have to ride by themselves in the freight elevator.[30]

Ruderman tells another revealing story:

> Once after work I went with some guys from the job to a bar. We were not filthy or anything [but] we were wearing construction clothes. It was a white-collar bar, and it was about 4 PM and we sat there having some beers. Then around 5 PM, the bartender said, "Excuse me,

I'm going to need those chairs." Apparently, the white-collar crowd was about to arrive. He did not want some blue-collar people sitting at the bar. He had no problem saying that. It wasn't like we were cursing or being rowdy. We just had to leave. Things like that really hurt the men.[31]

What consoles the men is knowing they can do the dangerous work that the white- and pink-collar workers who snub them could never do. Physical bravery is a crucial part of male construction workers' identity.[32] If women—who are perceived of as timid and fearful as men are brave—can do their work, then what remains of their self-respect?

Family Wages

The capacity to support an entire family with a single salary gives a male electrical construction worker a position of privilege inside the home.[33] When a woman appears on the job site and lays claim to tradesmen's work, it throws that whole dynamic out of whack. Ruderman explains why gender privilege in the family is important to the tradesman:

> Maybe there's a lawyer who lives down the street from them in the suburbs who makes about the same money—or even less—but he enjoys a status that the electrician, no matter how hard he works or how much money he earns, will never achieve. So, I guess he has to rest on the fact that, well, he is at least the "man of the house," and his work is really important. And when some woman comes along and does the same work for the same money, well, it cheapens him. That is what I think goes on.[34]

Men seek to exclude women from their trade to relieve the humiliation of class snobbery and because women in their trade threaten their compensatory gender privilege as "kings of the castle" at home. Women on the job then become their victims.

Both male and female construction workers have a special attachment to their work. Its physical challenge, its call on their skills, and its inherent danger endow workers with dignity and craft pride. This spirit bonds them to one another. The bonding is important: most electricians move from job site to job site, and don't stay together for long.[35] The construction workplace, which requires the integration of many craft skills performed in coordination, is threatened by any new worker, especially if that worker appears unpredictable. For male electricians, women, especially women with small children, are perceived as unpredictable. What the men see as a series of tests of a coworker's reliability can appear to the woman worker as unprovoked harassment.

For working-class men, equal treatment for women is unfamiliar territory.[36] The feminist revolution of the 1970s may have convinced many people that there should be equality in the workplace and in family life, and made feminist ideals part of mainstream American culture in the twenty-first century, but it had little impact on working-class communities.[37] Apprentice Emily Forman puts it this way: "I learned that men are really used to dealing with their wives or mothers, daughters or sisters, and they don't have too much experience dealing with women as coworkers or friends."[38] As a consequence, these men "try to place you as their wife by confiding in you about personal and emotional stories. Or the older guys treat you like a daughter: they take you under their wing—but patronize you. In general, when they can't place you as their mother, wife, daughter, sister, or whore, they get angry and confused."[39]

As in the broader working-class culture, masculine identity on the job site is defined in opposition to what is considered female.[40] Physically lighter jobs, such as handling snap-on electric boards (which require almost no skill), are called "tit jobs" and normally reserved for tradesmen over 55. Lax or careless male workers are called "girls" or "women," to express contempt for their failure to shoulder their share of the load. To call a man a woman means he is physically weak, unwilling to take risks, squeamish about unhealthy or unsafe working conditions, and generally too demanding.[41] This woman-baiting practice serves to discipline male construction workers, teaching them not to complain if they want to identify with the brotherhood. As in military training, intensifying the opposition between male and female is used to raise men's morale as a team while reinforcing conformity to the chain of command.

Women have their own special role in working-class blue-collar culture; they are the spiritual guardians of society who are protected at home while their husbands go out and work. Thomas Van Arsdale, who succeeded his father Harry as business manager of Local 3, puts it this way: "I believe in the traditional idea that there is a man and a wife and children, and the man goes out to work, and he should be able to get an income that is sufficient to support his wife and child without the wife having to work and the children being neglected."[42] By that reckoning, women who violate this pattern are downright harmful: "We're now paying the price for all those children whose mothers have had to go to work. Now, that's not to say that you don't have to have some provision for a woman who has lost her husband and has children. But she was married in the first place."[43]

The younger Van Arsdale was not, however, opposed to women without children working. "There are many women who choose not to have a family. These women should have equal opportunity in the workforce in whatever area they wish to pursue. They certainly should not be denied work simply because they are women."[44] Such formulations allow men in the brotherhood who think like him to maintain that they are not sexist.

Gender and Skill

Male construction workers tend to mystify the work they do as inaccessible and highly complex.[45] Handling power tools, reading blueprints, running cable, working with a powerful and dangerous energy source: all are considered as inherently masculine as a Y chromosome. Male electricians have enormous pride in what they do and consider themselves the professional equal of anyone. Veronica sums up this attitude well:

> If a guy said on the job that his son was going to be a doctor, the other guys wouldn't be impressed at all. They would look you square in the eye and ask you, "What's the difference between a doctor and an electrician? A doctor is a mechanic that works on people. He puts together and takes apart people. An electrician works on buildings. We both do the exact same thing. It's merely our patient that is different. . . . A doctor is just a glorified electrician that's making too much money."[46]

New technology like fiber optics has made electrical installation much simpler than before. This de-skilling threatens craft pride, the exclusivity of the skill, and the rationale for high wages of all electrical workers.[47] Women are blamed for this increasing degradation of the trade—as they seem to be blamed for any circumstance that threatens male esteem. Defensive resistance goes deeper than the desire to maintain an exclusive, skilled trade; it is tied to the male electrician's wish to maintain an exclusive, skilled, *male* trade. Ruderman explains:

> It's like I'm an affront to their masculinity. If I can lift the same pipe they can lift, or I can bend the same pipe they can bend, or if I can do a better job than they can—it's really a challenge to them. I mean, this is really a man's world and they are supposed to fit like a shoe. All of a sudden this woman comes in and the shoe fits her better or just as good. The men have this saying, "If it were easy, a woman could do it." Now, women are in there doing it, and the guys don't like it one bit.[48]

Anita, an African American journeywoman, notes that her male co-workers' anger toward tradeswomen stems largely from their fear that people might realize that there is not really so much heavy physical work to do on a construction site. Undermining the macho image of their work threatens their power and privilege.[49]

Gender and Ritual

Young apprentices and newly inducted mechanics become full-fledged union members by absorbing the customs and traditions of the brotherhood, which include hazing rituals.[50] Jokes and pranks are not just a way of breaking workday monotony but a method of testing the character of initiates. For instance, an apprentice might be asked to retrieve equipment that does not exist. How he reacts helps determine his status as a "regular" guy on the job. Humiliations can be used as punishments; pranks, teasing, and jokes can be used to harass a worker who is perceived as careless, whose behavior endangers the safety and health of coworkers, or who is not "pulling his load" on the job. The goal is often to harass the offender into another job placement.

Jokes, especially offensive ones, are part of how women, along with minority men, learn their place among the white men on a work crew: "The conversations and interactions on construction sites are mostly jokes," notes "Phil," a former carpentry journeyman. "And everybody has a joke which is either ethnic or about women." A woman's best defense is to trade insult for insult. It helps to be quick with the combative humor. Maura O'Grady, an Irish American journeywoman from a working-class family, relates the following incident:

> This guy was harassing me by "mooning" me on a ladder. Finally, I looked up and there was his butt right in my face. I mean—another step. I do not know how I thought of it, but I looked up and said, "Hey, Pete, why didn't you shave this morning?" Let me tell you, those guys threw themselves on the floor laughing, pointing to Pete. The poor guy lifts his pants and mopes down the ladder. But I had turned the tables. So everyone thought I was funny and great and said, "Oh boy, Pete, you thought you were going to get her, but she zinged you." The story must have gone to every construction site in the city.[51]

Her response helped O'Grady win acceptance in every work crew she joined afterwards.

Female electricians frequently remark on how often their male coworkers speak to them about very personal problems. Melinda, a Hispanic female construction worker says, "If you think Dear Abby has heard it—I think I've heard it all. The guys constantly talk to you about their girlfriends, kids, ailments and they think you have naturally broad shoulders. They will always say, 'Melinda, I have this problem and I have to talk to you.' They would talk to me much more than to the guys, especially about problems they were having with women in their lives."[52]

Women electricians also note that male workers like to talk to them about matters such as birth control, the women's movement, and how to raise children. Evan Ruderman observes that for some men, "working with women in a close partnership seemed to spark something sensitive and confidential in the men; they would never be so open with another male worker for fear of damaging the male act that they put on for each other." In this case, the presence of a woman on the job site even encouraged discussion and debate about broader issues, as Ruderman describes:

> There were five of us in a crew and I encouraged the guys to pick a weekly topic and everybody would read about it. Then we would have a debate. It was winter on this lousy construction site where more deaths had occurred than any other job I have been on. It was a really scary job. We used to hide at coffee time and get into these whole long discussions. It was really interesting to see the guys just wanting to talk about stuff that wasn't about the job and their family or complaining about their wives. In other words, they were eager just to learn and not be a part of this macho thing.[53]

Female electricians are bemused by how their male colleagues treat their wives. On the one hand, male workers insist that their wives should be the primary domestic caretakers of the family. On the other hand, these same men seem to resent the women's very dependent role. Apprentice Harriet Shroup says male apprentices "are really afraid of being 'used' by women. They seemed to have developed this incredible paranoia that they grew up with of being used by their mothers, sisters, and wives."[54]

Thanks to the family wage, many journeymen's wives are not compelled to seek outside employment. This domestic arrangement reinforces traditional stereotypes of masculinity as active and independent, and of femininity as passive and dependent. Paradoxically, the male worker's pride in being able to support his family can be accompanied by resentment of that family as parasitic. Nevertheless, male electricians have come to believe that their "traditional" family arrangement is the way every family should be. But their suspicions about how women use men carry over to the job.

Male apprentices fear that women will use their "flirtatious ways" to get "cleaner" assignments from foremen or that they will "use" male apprentices to do the undesirable work. Any display of sexuality (or sexual preference) can provoke mistrust or harassment. As one journeywoman noted, "Women are warned that the price for entering a man's craft and challenging their total dominance is that you have to lose your heterosexual feminine identity."[55] Male workers demand that women assimilate into masculine culture on the job, while, at the same time, they make that process difficult.

Male Fears

Of course, the reactions of union men to female workers are not uniform. They do, however, fall into patterns. Tradeswomen all note that the journeyman's age was a factor in their acceptance or rejection of female construction workers. Eighteen- to twenty-two-year-old apprentices are generally the most hostile, while electricians over fifty-five are most accepting. The ones in between—who often have large families—view women as a privileged group that might infringe on their ability to obtain overtime hours and their access to cleaner work. Journeymen in their forties are the ones who most readily perceive women as taking the place of a man who has to support a family.

Why are young male apprentices so hostile to the idea of female electricians? Individual male apprentices assert that while they might feel a sense of comradeship with female apprentices, they hate the idea of working under a woman journeywoman. Furthermore, many male apprentices feel that their buddies were not admitted into the union because women took their places—not on the basis of qualifications, but because federal affirmative action laws favor women over men, and this makes them angry. Some male apprentices even insist that they can accept the presence of women in such fields as engineering, law, medicine, or even management—just not in construction. Apprentices who went to public or other schools with women, as opposed to single-sex academies such as Catholic schools, or who worked with women before coming into the trade, were likelier to accept a female electrician, which suggests these attitudes can change.

For many male apprentices, the work ethic based on brotherhood and its exclusive claim to physical and technical skill is one of the attractions of construction work in the first place. For them, male gender solidarity and the family wage system are an attractive part of the package. Even today, family wage is not just a matter of abstract ideology for Local 3: during recessions, women are the first to be laid off, and then single men, then men with young children. The Local remains reluctant to admit that women workers might also be supporting themselves or a family, though of course this is often precisely the case. Dorothy, a journeywoman, remarks, "A person is entitled to work for a decent wage. That's the way I see it. I am single and I work with a lot of men who are single, and so I feel like there is a big hole in the buckets of their arguments."

Rational or not, the arguments for female inequality are vital to the preservation of the brotherhood. Craft construction work has little formal internal stratification of tasks (or wages), an aspect of the trade that craft unions struggle to maintain, and this precludes relegating women to less skilled, lower-paying work in the craft. Male workers and union leaders believe that when women earn the same wage as men, the employer is likelier to demand that men work for less. In addition, because in the end workers must make money for the boss, all employers will attempt to exploit all workers regardless of sex.[56]

Discourses on safety also play into the dynamic of gender discrimination. The need to justify high wages pressures workers to tolerate unsafe working conditions. Former apprentice Elvira Macri notes that the men think of themselves as similar to "firemen or policemen, in that it is real dangerous, and we deserve a decent living because we may not live long." "When a woman comes on the job," she continues, "now the guys need to either get rid of her or protect her. It's almost impossible for them to see her as an equal."[57]

Harriet Shroup, the former apprentice who dropped out of the industry and is now enrolled in premedical courses, remembers when she realized why male electricians were threatened by female workers:

> It was during a Christmas party given by the contractors on the job site. I finally met all the guys' wives and we had a long talk in the ladies room. What I realized was these guys needed to have these big Tarzan images at work so they could come home and tell these women, "Oh, honey, I worked so hard, I need a back rub, etc. . . . and all you do is stay home." Here they are going home and telling these women just like they were telling me on the job site that their wives could never do what they do and they are so tired they just have to rest. But here I was, a woman, doing the same work and going home to shop, cook, and care for my young son. When I met the wives, we were realizing these things, and it was a really big moment.[58]

Prestige and manly pride in this industry are defined through greater physical capability, greater earning power, and through a man's ability to dominate the women in his life, both financially and sexually. So while male apprentices and journeymen are wage laborers rather than proprietors, they constitute a gender-based aristocracy of labor. They like to see themselves as masters within the domestic sphere at home and in the workplace. "They [male electricians] don't want us on a financially equal footing," states one journeywoman. "They feel that it makes their role in a marriage obsolete because among white male electricians, there is the belief that every woman marries for money."[59]

Thus the mere presence of women on a construction site threatens concepts of masculinity, brotherhood, and gender hierarchy that are critical to the maintenance of male working-class identity. Such feelings run deep.[60] "When you top out [become journey level], then you can negotiate over scale with the contractors," observes a journeywoman. "They [the men] have given you their power. And now they are afraid of you."[61]

Employers

Female electricians also face hostility from their employers, who recognize macho culture as a method to spur workers to competitive feats of efficiency and strength. They also see manly camaraderie as a powerful force for social stability and self-discipline in the workplace.[62] Moreover, employers view women's circumstances such as pregnancy, child care, and domestic responsibilities as disruptions of the "rhythm" of construction work.[63] This penchant for disruption, they argue, disqualifies women from serious participation in highly skilled, high-wage work.[64] And since there is a close-knit statutory relationship between labor and management in construction (most of the electrical contractors are former Local 3 journeymen, and the Joint Industry Board [JIB] which governs the industry and workers is composed of Local union representatives and contractors), the boss's attitudes often intersect with the union's.[65] Management presses the union to deliver a cost-effective workforce, and male unionists treat tradeswomen's issues such as family leave, child care, and health and safety as impediments to that delivery, which in turn promotes hostility toward women workers. According to one male JIB staff member:

> If they hire a man, he comes in unless he gets sick; but if they hire a woman, they never know if she is going to get pregnant, and it is not a matter of one day. You know, she will be working, then she will be working with a belly and it's electrical work, and how can you let a woman like that on the job, you may have to pay worker compensation. After they have the child, they don't want them back for obvious reasons. They come back and a call comes in, the child is in the playground and has a fever. The woman takes off in the middle of the job. For example, they are putting up a fixture and she will drop the fixture and run. If you are a mother that is what you do. So this creates a problem for employers. So, they have good reasons for not wanting to hire women—economic reasons.[66]

Management is also likely to agree that women in construction do not have sufficient skill to perform the job adequately. The president of one multimillion-dollar contracting company that often handles small jobs in the city commented, "women do better on deck jobs because your skills don't have to show. Instead, you usually wind up going for coffee and you don't have to produce. But in the small shop, you are scrutinized, you have to be productive, and your skills show."[67]

Such attitudes deny the reality that much of the heavy physical labor and complex skilled work is being replaced by fiber optics, which require very little strength to install, and prefabricated circuit boards are assembled in factories, not on the construction site. That technology, remarks a former JIB staff member whose daughter is a journeywoman, "will remove a very important argument that you hear from contractors that women simply cannot do the work because it is too physical. Now they can't say that women can't lift those little wires, or clip in those little boards, can they?"[68] This de-skilling, and its attendant threat that increasing amounts of work can be performed by less-skilled nonunion labor, drives Local 3's leadership to exaggerate to contractors the degree to which the work is still difficult and dangerous.

When a woman does well, it is often ascribed by men to imagined sexual favors rather than ability. "My foreman on this job was really impressed with my work and my ability, so he wanted to promote me," recalls Colby Zieglar-Bonds, a mechanic of ten years' standing who recently worked on a prestigious job site. "But there has been a tremendous amount of rumors and harassment and jealousy from the guys. I feel like when I walk around the building, in every corner all kinds of ridiculous stories circulate about what is going on between me and the foreman. It annoys me, but that's the way it is, and that's way it's going to be."[69] Delores, a white journeywoman, likewise laments: "There is never a sense from the men I work with as a whole that I have gotten something because I was a good crafts worker." Some men are aware of this injustice. One astute and fair-minded journeyman observes: "If I get promoted to foreman and a friend of mine is the general foreman nobody is going to accuse me of screwing the boss. So, with women, there are a lot of extra twists."[70]

Sexually objectifying expression—pornographic displays, lewd language—is endemic and increases when a woman joins a work crew.[71] The objectification of women is even built into the nicknames of the tools and work processes themselves, such as the "nippling crew" on deck jobs (who disconnect short pieces of pipe sticking out of the newly poured concrete from the top of each floor) or the junction boxes on the sides of motors that are commonly called "peckerheads." (To avoid accusations of sexual harassment when women began coming into the business, some men changed the name of the latter to "niggerheads.")[72] Both male and female electrical workers, along with straw bosses and foremen, all attest to this phenomenon. In fact, it is commonly found in many, perhaps all, occupations where exclusively male domains are being invaded by women. When they were surveyed, over 75 percent of female firefighters said they had been sexually harassed in the form of unwanted sexual advances, requests for sexual favors, or physical contacts of a sexual nature.[73] Despite fire department policies against sexual harassment, when women reported it, only a third of the cases ended with the perpetrators disciplined; sometimes the women were themselves disciplined instead. Female firefighters were often advised just to ignore the harassment.[74] Offensive sexualized language and harassment is endemic in male-dominated occupations from restaurant kitchens to iron mines.

The widespread use of sexually explicit or vulgar terms for tools or processes makes it easy to try to embarrass women by making them repeat the offending word.[75] For example, one tool that can be used to bend pipes is called a "hickie." Journeywoman Maura O'Grady is constantly told, "Hey Maura, give me a hickie." She continues:

They have this tool which is properly called a drift pan. Sometimes, a pipe is crushed and you have to insert this instrument into it to try to uncrush the pipe and round it back out again. The men call this tool the "bull's dick." We also have diagonal pliers which are called "dykes." I always make a big point of calling them dykes because if I call them diagonal pliers, the guys will say, "Why, don't you want to call them dykes?" So, I just conform.[76]

Another misogynist affront on almost every job site is the commonly drawn caricature of one female journeywoman in the trade; she is called "mom."

The majority of men I have worked with have been children suspended in time who never quite worked through the trauma of their weaning period and hold serious grudges against "mom." She is drawn on walls in magic marker, pencil, and spray paint with her huge breasts and male anger in the form of penises [which] are drawn entering every cavity "mom" has.[77]

Among working-class men, shop talk, sexual boasting, and denigrating wives all promote camaraderie while avoiding any hint of homosexuality.[78] American blue-collar men have little tolerance for homosexuality among men or women. Sometimes, acting like one of the guys makes the woman "butch," subject to the accusation of being a lesbian (or, as the men sometimes put it, a member of the "nippling crew"). Ruderman believes that the obsession with sex "has to do with growing up with mothers who are really repressed and in families where sexuality was repressed. This is a kind of rebellion for them. I'm certain they can't act like this at home with their wives or parents, so they can really act out like this among each other on the job." In a version of the Madonna/whore dichotomy, for their wives, "then they can go home and be decent."[79] On the job the men must actively repress any affection they may feel for each other to avoid being thought of as gay. Ruderman notes that when she works closely with a man, it is only natural that he might develop warm feelings towards her.

And it might not even be sexual, it might just be affectionate, but it is hard for them to deal with that with me or even with another man. So sometimes, they get really rough. I mean, they are really physical, a lot of times beating each other up. Sometimes they even pretend like they are gay and start hugging each other—because they really want to hug each other, but the only way that they can do it is to make it a game. I think that is their way of saying they like each other.[80]

Horseplay among male workers harks back to boyhood camaraderie. It helps take the sting out of natural job disappointments such as a difficult work site or the lack of promotion. Mainly it supports the buddy system, the informal workmen's circle of mutual support, and the solidarity of brotherhood. Aggressive differentiation from all things feminine functions as a mechanism for gaining group acceptance as well as maintaining self-esteem.[81]

Fear of being thought of as "unmanly" on the job site readily expresses itself in verbal and physical actions. The behavior resembles that described by Christine Williams in her study of male recruits in the U.S. Marine Corps, who constantly define themselves in contradistinction to unmanly men.[82] Much of the language on the construction site is blatantly homophobic. Getting disciplined on a construction site is being "screwed up the ass." An unfair foreman "screws" or "rapes" his crew. Indeed, workers say they have been "raped" by him. As Lynne Segal, author of *Slow Motion: Changing Masculinities, Changing Men*, points out: "the association of poor or improper work performance with unmanly sexual behavior reinforces the notion that manliness is defined in opposition to femaleness, and almost defines away the possibility of accepting actual women as capable coworkers. Homophobic contempt for the 'feminine' in men generalizes as contempt for women." Of course, any man's suspicion or fear of evidence of femininity in himself must be expelled through histrionic rituals of hypermasculinity.[83]

Talk about women as sex objects is one such ritual, and it is actually encouraged by employers, who see pornography and sexual innuendo as harmless male pleasures and a distraction from poor working conditions. Contractors often distribute cheesecake calendars as part of Christmas bonuses or as goodwill advertising. The use of pornography as compensation for unpleasant work helps explain why, generally, the more miserable the working conditions, the more suggestive and pornographic the material displayed at work sites.

Women do object. The calendars were the inspiration for an angry letter to the union management from the Women Electricians of Local 3:

> We are dismayed to find that a Local-affiliated supply house was adding to the problem of sexual harassment by distributing pornographic calendars as Christmas gifts to our male coworkers. No woman should have to be subjected to the embarrassment of having one of these calendars hanging on the job. We strongly urge that all these calendars be removed from jobs and that supply houses are discouraged from printing them in the future.[84]

The Local's response proved disappointing. Many board members defended the possession and display of pornography on the job as a First Amendment right. One Local official responded, "I have not seen [the calendar] but suggest that you set out to solve this problem on a job-by-job basis where women members of the union are in an embarrassing position because of it. . . . Please realize that anything beyond that represents an intrusion into the private lives of our membership which is not the policy of the union."[85]

This response assumes that tradeswomen must conform to the construction site, not that male workers should accommodate women. "I worked with one woman who thought she could come in and change the environment," recalls Dominick, a middle-aged journeyman. "Women won't make it in the industry if they are too self-aware of being female." Of course openly displayed offensive pornography will make most women quite aware that they are female.[86]

Not all women react the same way to pornography and sexual harassment. Reactions are shaped by the woman's class, race, sexual preference, and educational background. The pioneer women who entered the construction trades—white, middle-class, often well-educated, feminist, and sometimes lesbian—viewed pornography as a form of unforgivable violence against women. They reacted accordingly. Beth Schulman, who is Jewish and was one of the first women to become a journeywoman, describes what she did to combat pornography:

> They told me to clean the men's locker room because they were going to have a Christmas party there. I was pretty pissed off that they had chosen me as a woman apprentice to clean the men's locker room. I figured I don't even use it, why do I have to go in there? But I did it anyway. When I went in, the place was plastered with porno pictures of women. So, I figured, well they told me to clean it and I considered those pictures filth. So I tore every single picture down, tore them to shreds, and threw them in the garbage. Then I cleaned the place really well.[87]

In contrast to the direct action typical of the pioneers, working-class women have been more concerned with the lack of training during their apprenticeship than with pornography. Personal economic success is more important to them than social justice. Black women in particular tend to prefer to deal with harassment incident by incident rather than as a pervasive and endemic problem. "I sat stoically through disgusting jokes and sleazy stories, all kinds of allusions to the female purpose and open viewing of pornography," recalls Shannon Spence, an African American journeywoman electrician. "There are many ways," she continues, "to handle uncomfortable situations that arise on the job."[88]

Fear of retaliation from contractors and the union have shaped responses to pornography and other offensive features of the job site: "I choose to deal with them constructively, while never endangering my reputation by being disrespectful or rude because you never know who your partners know in the upper echelons of the trade that could hold you back years with just a phone call."[89] In Shannon's opinion, white women separate themselves from their male coworkers, treating them as adversaries, and too often resort to formal grievance procedures. "Black women, we take care of the harassment on the spot. My thing has been if someone has tried to harass me, I will turn anything into a joke and at that point it's over for me."[90]

The African American women who become skilled electricians are generally poor and working-class, with little or no education beyond a high school diploma or GED. They usually support themselves and often also small children. These women expect that they will be permanently assigned to pollution control plants or deck jobs, where they will work hard but learn little. Many observe that while there is no formal internal stratification in the trade, women are more likely to be assigned to less-complicated tasks such as lighting, switches, piping, or receptacle inputs than to the more prestigious and complex work of building circuits or fire alarm systems, or reading blueprints. They hope that eventually knowledge and competence will earn those better assignments and more overtime.

Many of these women regard complaints about cheesecake calendars as counterproductive and trivial.[91] Joleen, an African American journeywoman, is philosophical about sexism in the workplace. "It's the way they treat women everywhere. And that goes for the Caucasians or anyone else. It's very hard for men to rid themselves of these attitudes. Instead, they carry them to the job." Black women may be just as incensed at harassment as their white counterparts, but they believe that women should try and work with men rather than confront them. Joleen advises, "Always conduct yourself in a manner that is complimentary to you and do not stoop to anyone's lower level, no matter how tempting or creatively satisfying putting him in his place may be."

One way to cope with these conditions is to develop a good relationship with the guys on the job before asking them as a personal courtesy to remove the offending picture or to engage in less direct sexual discourse. Joleen again: "Men like to work with women as partners because women will show them different ways to do things and generally won't be as stubborn or competitive as men." Once they have established some rapport with their partners, women can offer some suggestions. In any case, a woman who refuses the role of helpful partner risks being labeled uncooperative, feminist, or lesbian. "I was timid and afraid to stick up for myself for fear of being labeled the feminist bitch on the job," says Martha.

Women electricians do find that men react more intensely—whether positively or negatively—to women from their own racial or ethnic group. One Jewish tradeswoman remembers that a Jewish male coworker repeatedly said, "Hey, what are you doing here? You're Jewish!" expressing the common view that Jewish women should not be in construction work at all. Latinos are well known for their protective attitude toward Latinas on the job sites. African American men have complicated feelings about African American coworkers. They take racial pride in their coworkers' achievements, but they worry that white men will sexually haze black women, and that they will not be able to stop it. Yolanda, an African American journeywoman, observes:

> Black men are concerned that you do not let the white males think that you are an "easy get over" or that you are going to accept rude talk or attitudes. I have had problems with black guys who have come over to me and said, "Why do you let that white boy touch you?" Because you have to accept a white male touching you in a joking way, even though you may not like it.

As Shirley Merriman-Patton relates, black men also can be tough on black women on the job:

> Some of the brothers, especially those in their forties—this is where you really get it—they want you to be highly respected. They feel you are out on the job, you are out working and you are a business woman who is independent. They are looking at you to make sure that you carry yourself a certain way because you are, in their eyes, representing all black women, not only in the trade but outside in society as well.[92]

Merriman-Patton notes that middle-aged black men are especially concerned with how black women dress on the job, and how white men view their sexuality:

> I happened to be wearing a pair of old faded jeans that day and because of the lifting and bending I was doing on this job, my pants seam in the back opened and the white guys all became flustered and red. The black guys I was working with got very uptight and didn't know what to say. One guy finally stepped out and commented, "Do you know you have a hole in your pants?" I was very careful when I answered him, knowing that he was very sensitive. He was looking out because he felt that a white guy might be looking me over and this made him feel very offended. To some extent, I can understand it; after all, we are both black. But people get holes in their pants and guys are going to look. I was on the job and the hole was there. I couldn't do anything about it.[93]

Colby Zieglar-Bonds, a white journeywoman traveler, reports that black men invariably treat her better on the job than her white male coworkers. In her experience, black men do not act like they have to put another person down: "Black men have treated me very well but it's common knowledge that when I befriend a black man on the job, the white men will treat me like a whore. White men don't accept blacks generally, but some are more hostile toward their own because they feel that 'white women should be taken care of.' "[94]

Barriers to Justice

Because of the entrenched resistance to their presence, women electrical construction workers are reluctant to pursue their complaints through the established grievance process.[95] Because contractors and union officials are indifferent or hostile to women's concerns, the grievance process is inherently biased against women. This union failing exists across all the construction trades. Earline Fisher, an African American pioneer carpenter from New York City, has experienced dozens of layoffs while her white male colleagues are always working. "They talk about what they have, the house they have, the car and the boat, they have their little clique on the job and they are always working, it's the usual."[96] Her last layoff was from the Jacob Javits Convention Center in Manhattan, where she had been paired with an incompetent African American male carpenter. Earline explains: "He immediately tries to undermine me, and he's always talking about how he was an apprentice three times, and bragging that he never did anything on this job, you know, all negative stuff, nothing that you want to hear when you are working and people are watching you perform." There was only one bathroom on this job site. When Fisher complained to the shop steward, "He said the only thing he could do was to put a sign up there, and he didn't put a latch up or nothing, instead he said he would 'watch' for me."[97] But the next day, she was laid off, along with the African American male who was "not pulling his weight." Fisher believes that she was set up for failure by being paired with a known liability: "I was ranting and raving, I told them that you guys always have this shit going on, just putting people in for two or three days' work, and after that they drop you, especially women. I haven't had a full paycheck for two years—not one week's pay."[98]

After the layoff, Earline's black male colleague made what she viewed as "some sort" of deal with the shop steward and he was back on the job the very next day. Like many tradeswomen, she describes her attempts to grieve her layoff as futile:

> I went over to the union hall and complained that there was no shop steward on the Javits job. I told them that I needed to file a grievance for unlawful layoff and the guy in the union office got really nervous. I couldn't get anybody to file this grievance. Their lawyer was there and so was the business agent who could have taken care of it,

but everyone was avoiding me like the plague. Nobody would file this grievance, and no one would see me, and in retaliation they are not giving me any work.[99]

In the electricians' trade, there is clearly a double standard in the grievance process. Marian, an African American journeywoman electrician who has family in the industry and the union, observed that when a male electrician is called on the carpet for absenteeism, drinking on the job, fighting, or anything else that can slow the job down, he is brought before the Apprenticeship Council—distinct from the Apprentice Advisory Association which is the previously mentioned junior fraternity in the Local—which hears grievances and determines disciplinary actions for apprentices who breach work rules on job sites. Union officials and contractor representatives, many of whom went through the apprenticeship program, make up this committee. "There is usually someone to vouch for the apprentice, and he is put back to work with a warning," Marian, a member of the Amber Light Society observes. When a female apprentice has committed an infraction, she is sent to a different committee, composed of mostly male shop stewards and business agents, and she is usually laid off. Patricia, a female apprentice, was called down to the committee for disciplinary action and saved from this fate. Marian tells the story:

You know we only have one female shop steward in the entire construction division, and she got that appointment because I'm sure the union officials had to find a token or they would be in trouble with lawsuits. So this female shop steward out of ten shop stewards was on the committee and I called her and told her that Patricia was coming down and the whole thing was a set up to get her laid off. She convinced the men to give Patricia a second chance, and that's how she avoided a layoff or being kicked out of the apprentice program altogether.

Resentment of the double standard has fed attempts to build a caucus of tradeswomen beyond the local union level at the international levels in the IBEW.[100] These attempts may have helped some women navigate these inequalities, but the barriers to justice remain high. Again, Marian relates:

When I was placed on a job, it was an open iron structure and after being there a short time, I was assaulted by my foreman. I was so afraid that I might fall. I complained to the shop steward and he said go to the general foreman. Then I was told not to come to work and to gather the facts for the superintendent. Then I received a layoff and later a rotation to a pollution plant in Brooklyn where there were unhealthy conditions.

The contractors are especially insensitive and inflexible when it comes to pregnancy and maternity leave. One former high-ranking staff member of the JIB put it this way: "I have heard there is an effort now that women are in the industry to get pregnancy and maternity/paternity leaves. While the union won't object to that, they will emphasize wages and conditions more and have the attitude, 'Well, who wants them to have children? That's their problem, not the union's problem.'"[101] And men fully subscribe to this notion. The attitude is, "Well, she's knocked up, let her husband take care of her." Employers view the issue in the same way. The same official paraphrased contractors' attitudes toward pregnancy: "Look, she wants to work here, have a baby, and then come back—what's going on here? You know we're not against children, but we have to put up buildings. You know, we want to run a business. Let them have children, but we have contracts and clients."[102]

As much as they might prefer that women not work in construction, contractors do have to comply with federal regulations. "Contractors are even more strongly aware of the discrimination factor when it comes to women than the Local is," according to the JIB staff member. "For them, it is a cost problem, and they deal with it because they have to. As far as saying that they don't want women, they are not going to say that."[103] Even sexist contractors will tolerate women in order to avoid expensive lawsuits. The contractors only complain when the union sends them "stiffs"—unqualified workers. The Local has been known to send substandard tradesmen to contractors—usually men well enough connected through kinship, friendship, or membership in an influential union circle. However, unionized contractors find it increasingly difficult to compete with the rapidly increasing number of nonunion contractors, and will complain to the unions about substandard tradesmen even if they are "all in the family."

Black women carry a double burden: they are abused on the job site by white men and resented by black men, and are often the first to be laid off. Paradoxically, contractors prefer to recruit black women; hiring them gives contractors what are commonly dubbed "double bonus" points, helping them comply with government targets for both race and gender. Thus the number of black women in the trades grew faster than the number of white women in the tri-state greater New York area.[104] Regardless of their generally anti-women stance, the unions do try to help union contractors by providing workers who can help them meet the antidiscrimination requirements for bidding on government jobs.

The size of the contracting firm also helps to determine the degree of resistance that women experience on job sites. Large construction contractors working on federally funded projects, like the redevelopment of lower Manhattan, employ hundreds of workers and resist women less. In fact, they need to recruit women, at least in the initial stages of the bidding process. Unfortunately, these work sites often give women apprentices little chance at skilled training; the women may wind up going for coffee or cleaning out the temporary shanties where workers change and store work clothes and tools. Even the larger contractors who get billions of dollars of skyscraper work funded by federal subsidies and who feel compelled to show some good faith effort in hiring women may still oppose their presence on the job because they see women as sexually distracting to the men and, therefore, impediments to efficient production.

By all accounts, however, according to unionists, officers of the JIB, and women and men electricians, the greatest opposition to women on the job site comes from small contractors.[105] These employers focus on getting the job finished on time. Hostage to the high cost of loans and materials, they demand a high rate of productivity. They believe that women are less skilled. They consider women a disruptive force on the job site because of the "flirtation factor." They expect that women may have to take time off to care for a sick child, or may become pregnant, or need time off during their periods. They fear that women are just not up to the job. These attitudes reinforce gender discrimination.

Federal law requires private contractors to fulfill hiring goals for both women and minorities to be eligible for federal construction work subsidies on road, bridge, and transportation work. For unions, these federal requirements pose an additional risk. The growing open-shop movement and its powerful lobby, the Associated Builders and Contractors, Inc. (ABC), certainly understands that discrimination against tradeswomen is a source of vulnerability in craft union brotherhoods. Although the ABC has a worse track record than unionized construction, it is currently using this argument against the use of project labor agreements (PLAs) on federally funded work sites.[106] Nonunion employers and clearinghouses for nonunion tradesmen stand to benefit immediately and directly if craft union apprenticeship programs are decertified on the basis of racial or sexual discrimination by oversight agencies such as the New York State Apprenticeship Council. If the craft unions lost control over the apprenticeship system, prevailing wages would be undermined, and the unions themselves would lose their power.

5
Race for the Brotherhood:
The Ironies of Integration

We didn't know if we could crack open the trade. The Jews and the Italians did it and 1960s was our turn. Harlem Fightback was instrumental in our getting a foothold in the Local, and the leadership formed a social club named after Edison's protégé, Lewis Howard Latimer, a black electrical whiz.
—LOCAL 3 BLACK JOURNEYMAN PIONEER

L et us step back and ask some structural questions: Beyond the brotherhood ideology discussed in the previous chapters, how can we understand the continued exclusion of women from an industry which they have struggled for so many decades to enter? How does this struggle compare with those of other outsiders, such as minority men? Why has this historical pattern been reinforced more strongly in construction than in other industries? In this chapter, I take a closer look at questions of race—and at the puzzle of why and how, despite the strong degree of racism in society, male racial minorities have made some inroads into the electrical brotherhood (and in the building trades generally) while women, as writer Mary Wollstonecraft described for early modern fraternal guilds, remain "outside the door."[1]

The differing experiences of male racial minorities and women in the electrical trade and brotherhood are a microcosm of how the subtle institutional aspects of race and gender relations structure the dance between men and women in construction jobs. Their experiences also provide an opportunity to examine the influence of the social movements of the 1960s and 1970s for equal opportunity in employment, as well as the impact of their decline in the 1980s.[2] The experiences of women and minority men prove that it is much more than prejudice that impedes women's way, and suggest that the Sisyphean task of changing culture is not the only possible route to achieving redress. Aggressive governmental action, legal challenges, and the pressure of public opinion remain potential avenues for change.

The Theoretical Context

In *Gender Relations at Work*, a now-classic account of gender integration in the auto industry, Ruth Milkman demonstrated that auto industry employers after World War II dismissed wartime female workers in favor of returning veteran male workers, despite the fact that the wages of the women were lower.[3] This gender preference reflected historical assumptions about the workplace generally, and gender-specific customs and traditions of the auto industry and its union, the United Auto Workers (UAW). Milkman made a broad-based critique of much of the literature on race-and-sex segregation in the labor market. The dominant focus in that literature continues to be on historic and economic skills mismatch and human capital theories, both of which stress debilities on the labor supply side in the form of inadequate training and education, and wage competition theories, which study how much it costs for employers to employ certain groups.[4] For Milkman, these theories must be tempered by demand-side theories of how specific industries, in specific historical contexts, evolved to prefer laborers of one sort over another. These realities, she points out, can often only be explained through fine-grained study of historical contingencies.[5]

Barbara Reskin and Patricia Roos, in their recent study *Job Queues, Gender Queues*, conclude that women enter male occupations due to an undersupply of male workers resulting from rapidly expanding growth in the service sector.[6] Other factors, such as the need for women employees in order to comply with antidiscrimination legislation, raise costs for employers who persist in their preference for male workers. Reskin and Roos argue that employers are willing to integrate women into male-dominated fields such as accounting and systems analysis in order to lower costs (by avoiding costly lawsuits), but this does not automatically lead to economic equity with men.[7]

"Queuing theory" does little to explain the dynamics of integrating women into craft union work; Milkman's approach is more appropriate. Wage differentials between races and genders are irrelevant; equity in skills is vouchsafed by an intensive apprentice system; and access to employment is determined not so much by pre-existing skills and education but instead, quite explicitly and literally, by "who you know."[8] Further, nothing in any theories of skills mismatch, human capital, wage competition, or queuing can explain an irony in Local 3's history: why the union was remarkably progressive in taking political leadership among craft unions in the 1960s and 1970s to induct minority men, yet very reactionary in the absorption of women.

Racial segregation in the building trades was infamous. Building trades unions had historically excluded minority workers outright or established segregated locals. In 1960, blacks represented just 2 percent of all union apprentices in the United States. "Exclusion in the craft unions is so complete," the director of the Chicago Urban League testified before a congressional committee, "that segregation would be a step forward."[9] The building trades unions were so recalcitrant on the issue of integration that Whitney Young, executive director of the Urban League from 1961 until his untimely death in 1971, likened the entrance of blacks into the building trade craft union apprenticeships to "Apollo landing on the moon."[10]

The modern apprenticeship program was created by the little-known 1937 Fitzgerald Act, which established federal and state standards of certification.[11] The act gave states the authority to establish their own "hometown" apprenticeship councils that would regulate certification for unionized apprenticeship programs. These agencies came under their respective state labor departments, which in turn reported compliance at the state level to the federal Bureau of Apprenticeship and Training in the U.S. Department of Labor.[12] The New York State Apprenticeship Council is the regulatory agency overseeing apprenticeship program certification in New York, and the New York State Labor Department is ultimately responsible for overseeing the standards and requirements of apprenticeship.[13] In the 1960s, bowing to pressure from courts, federal regulators, demonstrators, and legislators, local apprenticeship councils in over thirty states were required to file annual reports to document compliance with the goals and timetables of affirmative action programs to increase the number of minorities working for federal contractors and subcontractors in the construction industry.[14]

Some intractable unions fought legal battles to keep blacks and other minority male workers out. These efforts proved expensive, time-consuming, and ultimately futile. Their failure encouraged other local building trades unions, such as the IBEW in Seattle, Washington, to work cooperatively with community outreach and training organizations and with the Department of Labor.[15]

In 1969, the Building and Construction Trades Department of the American Federation of Labor-Congress of Industrial Organizations (AFL-CIO) endorsed the concept of affirmative action in apprenticeship.[16] Verbal declarations alone hardly satisfied black civil rights activists such as Adam Clayton Powell, the first black congressman from the Northeast; Roy Wilkins, the head of the National Association for the Advancement of Colored People (NAACP); and Herbert Hill, the NAACP's National Labor Director. Hill testified on behalf of the NAACP before a legislative committee in Washington, DC, on the lack of progress for inclusion of minorities:

Spokesmen for organized labor tell us that the ratio of non-whites among registered apprentices increased between 1960 and 1970 from 2 percent to 7.2 percent nationally. But, the significant ratio is the number of Blacks and other nonwhites in the skilled trades measured against their total number in the community at large. In relation to this ratio, the Outreach Programs [funded by the federal government] have only succeeded in maintaining the appallingly low levels of non-white participation reached in the early 1960s.[17]

By 1979, minority men nationally comprised 17 percent of all registered apprentices in the construction trades (compared to less than 6 percent in 1967). Much of this progress was a result of the actions of black civil rights organizations, such as legal challenges brought by the NAACP, direct efforts by the Congress on Racial Equality (CORE), and the Urban League's minority training and outreach programs. By 1980, these programs, funded nationally by the Department of Labor, were established in 114 locations and had helped recruit more than forty-five thousand minority apprentices into the construction trades.[18]

Women and Minorities in Construction in the Tri-State Area

As a result of civil rights struggles, legal challenges, and direct action on the grassroots level, black and Hispanic men have made inroads into the building trades. Labor participation rates from 1970 to 2000 in the tri-state (New York, New Jersey, and Connecticut) area show a slow but steady advance in the employment of racial minorities (see Appendix C, Table C.2). African Americans have more than doubled their share of employment in the constructions trades, from 4.9 percent in 1970 to 12.1 percent in 2000. Hispanics more than tripled their share in a much shorter time, from 5.3 percent in 1980 to 18.1 percent in 2000. Similar advances can be observed among Asians and other ethnic groups as well, even though their numbers are still relatively small (2.4 percent as of 2000) (see Appendix C, Table C.2).[19]

It is important to stress that these advances in minority employment are mainly among male workers; the participation rates of female construction workers have only marginally increased, from 1.2 percent in 1970 to 2 percent in 2000 (see Appendix C, Table C.1). Industry-wide these are insignificant numbers. This pattern of exclusion has persisted despite the fact that advocacy organizations charged with recruiting women for trades work have reported a growing interest for jobs in those areas, especially on the part of poor and low-income women of color.[20]

Over the past four decades, African American, Hispanic, and Asian males have made undeniable progress in getting and holding skilled construction jobs. Despite persistent cultural, racial, and ethnic rivalries, men of all colors and cultural backgrounds have been able to transcend the barriers of race and ethnicity enough to bond together as brothers, buddies, and teammates in a work culture that demands such collective assimilation. Women are still "outside the door."

Emerging Trends Across Gender, Race, and Ethnicity

Even within the context of the marginal improvements in women's employment in the construction trades, there are differences in developments along gender, racial, and ethnic lines (see Appendix C, Tables C.3a and b). While the overall employment of white and Hispanic women has hardly changed over the past four decades, African American women have increased their rate of participation. Black women have more than doubled their rate of participation within the black population working in the building trades (from 1.9 percent in 1970 to 4.1 percent in 2000), while women's share of white participation increased only slightly (from 1.2 percent to 1.7 percent). The women's share of Hispanic participation actually dropped (from 4.2 percent to 1.6 percent between 1980 and 2000).

As for their employment in the construction trades overall, black women made up the smallest portion of the labor force in 1980, but the increase in their relative share has been the largest among women—their numbers show a fivefold increase from .10 percent in 1970 to .51 percent in 2000. Nonetheless, black women are grossly underrepresented in construction compared to their civilian labor force participation (7.7 percent) in New York.[21]

The participation rates of Hispanic men have also significantly increased. This may be due to factors such as the growth of Hispanic representation in the population as a whole, as well as among contractors and subcontractors, and the reported existence of an extensive informal referral network for jobs among Hispanic families. These increases in minority participation have meant a decrease in participation for white men, which fell from 93.91 percent in 1970 to 67.92 percent in 2000, a figure somewhat more in line with the proportion of white men in the general population of the New York City area.

Among women in the overall construction trades, being black appears to confer something of a relative hiring advantage despite the fact that the participation rate of black women is still very marginal.[22] Reasons for this advantage can be found on both the demand and supply sides. On the demand side, black women count for more "bonus points" (as both a woman and a minority) by the Office of Federal Contract Compliance (OFCC) for contractors who are competing for federally funded job projects, such as the rebuilding of lower Manhattan. On the supply side, black women, who have had historically higher labor participation rates than white women, have a strong incentive to apply for highly remunerated unionized construction jobs because they are more likely than white women to be single heads of households, and less likely to have the means for higher education. High pay, along with cradle-to-grave union benefits such as health coverage, pension plans, and college tuition, can transition low-income black women workers into the middle class.

The increased participation of minority men, especially Hispanic men, in the trades suggests that the use of family connections, which maintained the industry's whiteness for decades, is changing. Hispanics are more likely than blacks to own their own small contracting firms, so they are well-positioned to bring members of their family networks into construction work. Electrical construction work, where many contractors started as union apprentices and family connections were long the dominant mode of recruitment, may be especially supportive of expanding the Hispanic foothold in the industry.[23]

Despite the addition of women to the family networks of potential apprentices, the industry has not drawn in women as successfully as it has minorities. Nor has it transformed the brotherhood's fraternal culture. Despite all the litigation, federal compliance standards, advocacy, and outreach, the building trades still remain a fortress impregnable to women—black women, Hispanic women, Asian women, and white women alike.

Emerging Trends in Electrical Construction

Trends in participation by women and minorities in New York State's electrical and carpentry trades mirror the general trends in the construction industry for the tri-state area (see Appendix C, Tables C.4a and b). The data (which exclude apprentices) show women overall are only marginally represented (1 to 2.6 percent) among both electricians and carpenters. Moreover, the slight gains women achieved in the 1980s quickly tapered off in the 1990s. The participation rate of female electricians in New York State slightly increased from 2.53 percent in 1980 to 2.61 percent in 1990 and then dropped back to 2.5 percent by 2000, below even the 1980 level (see Appendix C, Table C.4a). Female carpenters increased their participation rate from 1.02 percent in 1980 to 1.57 percent in 1990, then dropped to 1.5 percent by 2000 (see Appendix C, Table C.4b).[24]

Perhaps these trends reflect economic changes in the industry. While the 1980s were a decade of dynamic expansion for both trades—15 percent growth for electricians and 43 percent for carpenters—the 1990s were a decade of contraction—employment down 6 percent for electricians and 19 percent for carpenters. Yet the change in rates of participation by women over these decades does not even come close to these swings. By contrast, minority men as a total have experienced a steady growth in both trades despite the ups and downs of the economic cycle. For example, the total minority participation rate of electricians increased from 15.3 percent in 1980 to 20 percent in 1990, and 24 percent in 2000 (see Appendix C, Table C.4a).[25] Similarly, among carpenters, the minority participation rate increased from 12.7 percent in 1980 to 18.4 percent in 1990, and 25.4 percent in 2000, and comparable to the findings in the larger tri-state area, Hispanic men have fared better than their black counterparts (see Appendix C, Table C.4b).

Men of Color: The Local 3 Story

Ultimately, the economics of the construction industry do little to explain the advances of male racial minorities in the trades. Though one might think that men of color would have a better chance of getting work in periods of expansion, the data show that economic expansions and contractions in the industry have had no significant impact. Black men in particular, though they first came into the industry in the expansionary early 1960s, made their most significant advances during one of the industry's most severe recessions in the mid- to late-1970s, and during the economic building boom of the mid- to late-1980s, black men actually decreased their share among new minority apprentice registrants. According to researcher Thomas Bailey, "the percentage of black apprentices in the New York/New Jersey Region fell by one-half of one percent between 1980 and 1986, while in the stagnation period of 1974 to 1980, their numbers rose by 3.6 percent."[26] Although labor participation rates for minority male electricians overall slowly but steadily increased over the last three decades of the last century (as illustrated by the 2000 census), status compliance reports for the electrical workers' apprenticeship program show a decline in the minority male apprentice share for new registrants from 27.7 percent in 1985 to 15.3 percent as of May 1989—a period when construction was booming.[27]

The story of the integration of minority men into Local 3 begins when the first minority men (primarily black and Puerto Rican) were voluntarily admitted to the craft division of Local 3 during the early 1960s. This crack widened throughout the late 1960s and 1970s via several pressure points. The civil rights movement focused on good-paying jobs in the northeast construction trades. Local union officers took political leadership in integrating minority men into the trade. Unionized contractors demanded referrals of cheaper apprentice labor (as opposed to journeymen) from union hiring halls. And community groups took militant direct actions.[28]

The year 1963 was pivotal for the induction of male racial minorities into electrical construction. Ironically, the momentum to penetrate the northeastern building trades was marshaled in the Jim Crow South. That year the brutal police suppression of demonstrations by black students in Birmingham, Alabama, was widely reported and televised. White apathy about black civil rights became sympathy and support. Civil rights leaders were able to increase pressure on President John F. Kennedy to take action against racial discrimination.[29] The Birmingham protests also galvanized organizing in black communities nationwide, and attention focused on addressing the direct causes of black poverty. After Birmingham, the movement attacked institutional barriers to good-paying jobs for blacks, specifically in two industries: manufacturing in the south and construction in the northeast.[30] With jobs like these white males with little or no higher education had been able to make it into the middle class. Civil rights leaders believed that access to these jobs would do the same for poor blacks.[31]

In construction, employers' practices were not the only barrier to racial inclusion. Building trades unions, steeped in nepotistic referral practices and unflinching assumptions of white male privilege, had long proved impenetrable to black and other minority men. The unprecedented economic growth in the 1950s and 1960s had bolstered the collective bargaining power of building trades unions, as well as other craft and industrial unions, contributing to their political ascendancy and influence on mainstream political parties.[32]

As a powerful political constituency for both Republican and Democratic parties, building trades unions especially influenced the actions of governmental regulatory agencies, including those charged with enforcing desegregation orders, such as the OFCC (originally established by President Harry S. Truman in 1951) and the U.S. Department of Labor's Bureau of Apprenticeship and Training.[33]

But desegregation orders were not the only challenges faced by employers, and unionists and building trades unions were not the only problem. At the time of the merger between the American Federation of Labor (AFL) and Congress of Industrial Organizations (CIO), six labor organizations still had constitutions that formally excluded blacks, Mexicans, and other men of color. They were the Brotherhood of Locomotive Engineers (BLE); the Brotherhood of Locomotive Firemen and Enginemen (BLFE); the Postal Transport Association, AFL; the Brotherhood of Railway Trainmen (BRT); the Order of Railway Conductors (ORC); and the Brotherhood of Latherers, AFL.[34] Eventually, these organizations removed their race bars but their battle for the right to discriminate against black men continued into the 1970s and beyond. Similarly, some building trades unions, like Local 28 and Local 86 of the ironworkers, decided to stay the course on their union discriminatory practices and went kicking and screaming to federal court over compliance through several years of litigation.[35]

In the early 1960s, new legislative codes recommended by the Kennedy administration that set hiring quotas for minorities and threatened to slow the construction of skyscrapers in big cities like New York if those requirements were not met brought on something of a panicked response from contractors and Local 3 to fill apprentice places with minority men as quickly as possible. This occurred at a time when many sons of white journeymen were removing themselves from the apprenticeship pool by pursuing new educational opportunities to enter white-collar professions.[36]

At the same time that the federal government stepped up efforts to ameliorate the effects of racial discrimination in the building trades, the unions were coming under fire from business interests such as the National Contractors' Association (NCA) and the National Association of Manufacturers (NAM).[37] Their criticisms were widely publicized. A *Fortune* magazine article described the building trades unions' "murderous bargaining strength" as the cause of spiraling wage-price inflation. The article also called for the end of union-run apprenticeships and for "new legislation against discrimination."[38] Proving that politics make for odd bedfellows, business interests and civil rights leaders demanded more recruitment of blacks into construction apprenticeships, for different reasons: the former to curtail control of the labor supply by the unions which, according to employers, were "the most powerful oligopoly in the American economy today"; the latter to promote black access to good-paying jobs.[39]

Local 3 business manager Harry Van Arsdale Jr. feared that public opinion, reactionary business interests, and politicians would weaken the union's control over apprenticeship, even as escalating costs associated with rising electrical journeymen's wages could price unionized contractors out of the market. He set out to secure a cheaper skilled labor force, and struck a deal with the president: in exchange for Kennedy's passing new apprenticeship legislation lifting the ban on contractors' use of apprentices on skyscraper work, Local 3 would allow new apprentices into the electrical apprenticeship program from outside the traditional father to son[40] sponsorship. This would be a first not only for the electrical brotherhood but for all the building trades unions.[41] Out of this new class, Van Arsdale guaranteed Kennedy that at least one-third would be minority men, thus setting a precedent among other building trades unions and a goal for desegregation in the industry.[42]

Van Arsdale expected opposition from electrical journeymen interested in keeping their trade "all in the family" and fearing increased unemployment due to automation. To mitigate this, he staged an eight-day walkout of 9,000 electricians in January 1962 against the National Electrical Contractors Association. Van Arsdale emerged as a national figure by settling the strike and winning a twenty-five-hour work week for Local 3 journeymen, with the fifth and sixth hours each day paid at time and a half, which earned him the *New York Daily News* title "The Big Boss of the Short Day."[43] Van Arsdale, insisting that reducing the number of hours for tradesmen was one way to cure the ills of unemployment among New York's electricians, invited the wrath of New York's builders who, according to an article in the the the *New York Post* on May 12, 1962, "were aghast at the cost of his [Van Arsdale's] therapy." On the supply side, Van Arsdale promised the Local would hold down overtime by adjusting its work rules, staggering hours, and admitting twice as many apprentices for the 1963 class, ensuring contractors a skilled labor supply for less than journeyman cost. On the demand side, he promised that the brotherhood would "make every effort to assure substantial minority representation"[44] for contractors required to comply with new regulations for minority workers on federally funded projects.

Everyone got something. Even though Kennedy felt compelled to publicly criticize Van Arsdale's "shorter work week" strategy, Kennedy appeared effective in getting the recalcitrant building trades unions to admit racial minorities. Electricians saw their work day shortened for the first time since the New Deal, from thirty-five to twenty-five hours per week. Building costs were contained to keep contractors competitive against nonunionized shops. Anti-union public opinion fueled by business interests and their allies in government was countered. For the first time the father-to-son membership limitation in the construction division, Division A, was relaxed. The union rolls opened up to blacks and Puerto Ricans outside the brotherhood's minority members who were in lower-paid divisions. According to an article in the *New York Times,* letters of recruitment were sent out to the NAACP, the Urban League, and other New York unions and employers, according to a 1962 *New York Times Magazine* article, urged them to make nominations without concern for race or religion. And Van Arsdale got a stronger union. In one fell swoop, he had addressed segregation in the industry and seized a vanguard political position as a progressive unionist. Of the 1,023 apprentices in the 1963 class, 240 were black men and 60 were Puerto Rican. (The history of this class is poorly documented, but by 1965 a hundred men, half of them black, had left the program.) In 1966, almost 200 apprentices became journeymen; seven were reported to be black.[45]

Van Arsdale went a step further. Infuriating both rank-and-file electricians and other building trades union leaders, he launched a broad public relations campaign and made alliances with civil rights leaders.[46] When Local 3 was hosting the Reverend Martin Luther King Jr. and Congressman Adam Clayton Powell Jr. at Electchester Brotherhood Communion Breakfasts, most of the other New York building trades unions —ironworkers, steamfitters, latherers, and plumbers—were in federal court facing legal challenges for their failure to comply with new affirmative action laws.[47] Among some of these unions racial exclusion was not only customary but a constitutional requirement. Steamfitters' Local 28, for example, faced lawsuits challenging its seventy-eight year practice to "automatically exclude" blacks from its constitution.[48]

Recruiting Men of Color into Apprenticeship

Harry Van Arsdale put his high standing and respect in the union (and in the New York labor movement generally) behind a new principle in apprenticeship recruitment: each newly indentured class of electrical apprentice inductees would be one-third African American, one-third Hispanic, and one-third white. However, status reports from compliance reviews show that the minority share of new apprentices fell short of these goals. For example, the

reports reveal that from 1968 to 1971 the number of minority males inducted into the apprenticeship program increased from 84 to 110, but fell
far short of even one-third of the 518 and 569 new registrants, respectively,
in 1970 and 1971.[49] By 1973, the Local could claim only 288 male minorities out of a total of 1,927 enrolled in the apprenticeship program.[50] Although he was unable to achieve his integration goals, Van Arsdale's actions,
daring, and vision compare quite favorably with the response of most other
leaders in related building craft unions.

The induction of that first class of minority men in 1963 and subsequent
attempts to integrate electrical apprenticeship through the 1960s and 1970s
owed much to the sheer force of Van Arsdale's will.[51] When civil rights authorities began demanding that the application process for apprenticeship
be opened to outsiders, the Local could have taken this as an outright threat
to its very existence, since the kinship recruiting system automatically provided mentorship and disciplinary control to young white male apprentices
by the relatives who sponsored them. The union sought instead to try to
preserve those values of discipleship by engaging the very organizations that
were challenging racial exclusion—NAACP, the Urban League, the Workers' Defense League (WDL), the Recruitment and Training Program (RTP),
the Negro American Labor Council (NALC), and other minority male organizations, including even a Local 3–sponsored center-city Boy Scout troop.
These organizations helped the union by recruiting minority men for apprenticeship, watching over them, and acting as a kind of surrogate kin within
the codes of the brotherhood system.[52]

One of the first and most important outreach and training programs
that facilitated the Local's efforts was the WDL, a human rights organization affiliated with the A. Philip Randolph Institute.[53] Ernest Green, one of
the nine students of Little Rock, Arkansas, Central High School integration
fame, headed the organization. Although Green's brother successfully sued
the steamfitters' New York Local 28 to obtain entry into its apprenticeship
program, the WDL's mission aimed to integrate black and minority men
into the building trades through the front door. According to Ray Marshall,
the former U.S. Secretary of Labor:

> The Workers' Defense League, which has devoted considerable at-
> tention to efforts to increase nonwhite participation in apprentice-
> ship programs, reports that it has had excellent co-operation with
> the [electricians'] union in general and with Van Arsdale in particu-
> lar. During 1965, in fact, the League reports that the local accepted
> thirty-five minority applicants recruited by the WDL.[54]

In the early 1970s, the WDL—nearly dismantled due to strife among internal
factions—joined forces with the RTP, another storefront agency with offices

in Harlem. Funded by the federal Comprehensive Employment and Training Act (CETA),[55] the RTP recruited candidates with or without high school diplomas, prepared those without diplomas to take the GED, and helped applicants to building trades apprentice programs complete the forms and gather the documents required for admission. In addition, the RTP offered math and reading instruction and practice tests for the apprenticeship exam, and invited speakers from various craft trade unions in the city to speak with potential applicants about the terms and responsibilities of apprenticeship.[56] Unfortunately, federal budget cuts in the 1980s dealt a devastating blow to the RTP and to the labor supply of minority men to the electrical brotherhood and to other unionized apprenticeship programs generally.[57]

Legitimizing Integration: Fraternalism and Direct Action

Van Arsdale faced opposition to racial integration in construction not only from other building trade leaders but also from his own rank and file. By reducing the work week for journeymen and raising their wage scales, Van Arsdale legitimized the entrance of racial minorities into the coveted apprenticeship. He confronted resistant journeymen by relying on the rhetoric of fraternalism and brotherhood: a commitment to bringing racial minorities into the trade, but only through the time-honored path of apprenticeship. By framing integration in this manner, Van Arsdale made the induction of male racial minorities seem simply a matter of fair ethnic representation within the long-established fraternal club tradition of the electrical brotherhood.

The legitimization process was not limited to recruiting efforts. Van Arsdale founded a new racial minority club, the Latimer Society, and redirected and elevated the role of an established Hispanic club, the Santiago Iglesias Club, from strictly servicing the needs of Puerto Ricans in the electrical manufacturing division to more broadly recruiting Hispanic men into the elite Division 1.[58] Thus he incorporated the club function into a union strategy to build solidarity across a culturally diverse membership. As a result of efforts like these, beginning with the admittance of that 1963 class, the entry of minority men into the union never had the same smell of outside interference that accompanied later efforts by agencies to pressure Local 3 to accept women.

The "official story" in the brotherhood regarding the Class of 1963, as told by union members, centers mainly on Van Arsdale and his cult of personality. Stan Brown, a former organizer in New York City's motion picture and production companies and an African American journeyman whose father was an electrician in Harlem Fightback, describes the "popular fiction" in the Local like this: "Van Arsdale was a far reaching and progressive thinker who wanted to bring the organized labor movement into the twenty-first century."[59] But the real story, Brown believes, especially about that first minority class had less to do with Van Arsdale's personal force of will than by the force of outside political pressure: "Van Arsdale traveled with Adam Clayton Powell Jr., recognized the strong political clout of Powell and other civil rights leaders, and believed that blacks coming into the union was inevitable."[60] Brown also thinks that Van Arsdale knew that women would also have to be accepted. "He knew it was going to happen but he wanted it to happen on his own terms." [61]

Even within the inner circle of the brotherhood, union leaders acknowledge that idealism was not Van Arsdale's only motivation. The "unofficial story," according to a seasoned journeyman, acknowledged the Local's fear of growing African American militancy on Manhattan building projects which fostered significant challenges to the maintenance of unionized electrical work in New York City. Disruption of work flow and schedules on construction jobs escalate contractors' costs. If the delivery of materials is either halted or slowed, or expensive equipment is damaged, shift changes for the various tradesmen are disrupted, schedules are delayed, building loans expire, and interest rates increase—all severely jeopardizing union contractors' ability to compete with the nonunion sector.

Lafayette "Buddy" Jackson, a former director of the apprenticeship program and one of the rare black journeymen whose membership dates from the union's very limited racial outreach efforts in the immediate postwar period, suspects that the 1963 decision to buck the strong backlash from Local 3 members and leaders of other building trades unions by voluntarily recruiting minority men into the apprentice program resulted simultaneously from adverse public opinion, direct action by black leaders and workers, and fear of potential violence on construction sites. He recalls: "In what was called the turbulent sixties you started having the civil rights movement because minorities were not really recognized and they were shut out of the mainstream as far as craft union jobs were concerned. But the Local was always way ahead and they could see what was coming, so they opened the apprenticeship to minority men."[62]

The demonstrations organized by leaders such as A. Phillip Randolph, the legendary black labor leader of the AFL-CIO–affiliated Brotherhood of Sleeping Car Porters, and by other militant grassroots groups such as Harlem Fightback, caused particular concern. Jim Haughton, a former associate of Randolph and founder of Harlem Fightback, recalls:

> Unemployment was bad in 1964, as it is now. I was working in the NALC with transit workers. And I had a group of longshoreman who were fighting the battle on the docks, against a whole group of white guys in that Local 1814 of the International Longshoremen's Association (ILA). We had utility workers, garment workers, and we

created an executive board and agitated on the question of employment, especially in the building trades. Workers then had the desire and motivation to protest. They were spirited, and there were some linkages to the Communist Party. They were very class conscious, and they had formed coalitions with white workers in the industry, and it created a mushrooming of a real movement in Fightback.[63]

Outdoor construction sites have always been vulnerable to protests, work stoppages by outsiders, sabotage, or outright physical attack, something that labor organizers have sometimes used to their advantage and that organized crime uses still.[64] Haughton remembers: "There were good, solid protests at construction sites. We became active in the antiwar movement. We got involved in public housing and picketed the homes of congresspeople. There was a whole ambiance of movement and spirit."[65] Surely it was better to absorb black males than to confront them, but the movement for inclusion of blacks into the trades resonated among other racial minorities. Militant Hispanic groups such as The United Bridge, organized by Latino craftsmen, picketed city and federally funded construction sites based in Hispanic communities (e.g., Harlem Hospital and the City University of New York), pressuring contractors to hire them and unions to sign them up.[66]

Civil rights leaders like Randolph and labor-based groups like the NALC and the WDL recognized the importance of getting white working-class men to support affirmative action programs, so they worked closely with Van Arsdale and Local 3 beginning in 1963 and continuing throughout the 1960s and 1970s. The NALC sponsored one hundred black men from the council into Local 3's apprenticeship program.[67] According to Haughton, "these tradesmen became the first members of the Latimer Society, and the core group that opened up the Local to black men."[68]

The Latimer society had its antecedents in a long history of struggle waged by successive waves of immigrant groups fighting to secure their own occupational niches and ethnic enclaves in construction.[69] Jewish craftsmen of the 1920s organized associations and fraternal clubs to break into the Irish-controlled New York City electrical trade.[70] Black electricians in the 1950s struggled to get well-paid craft jobs not only in construction but as electricians in the city's unionized motion picture industry.[71] Compared to the struggle of immigrant groups, the black struggle has at least one striking similarity and one distinct difference. Once blacks secured a toehold in the trade, they sponsored male kin into the brotherhood like the Jews and earlier ethnic groups had for the purpose of increasing their representation in the trade.[72] In contrast to immigrant groups, however, blacks were unable to establish themselves as contractors or subcontractors in the industry.[73] Observing this failure, some critics of the Local have derided the Latimer Society as ineffective and as a mere ploy by the union to create the impression that it was doing the right thing. From this perspective, shared by

Haughton and other more militant electricians in Fightback, "they [the black male electricians] established a 'beachhead' that the Local controlled."[74]

Nonetheless, Randolph's sponsorship of blacks into the brotherhood encouraged optimism among civil rights activists.[75] It also created a severe backlash. In 1964, Randolph and the NALC were confronted with a big rift across AFL building trades unions over the unions' outright refusal to induct minorities.[76] As head of the AFL affiliate Sleeping Car Porters' brotherhood, Randolph answered directly to AFL-CIO President George Meany.[77] Meany himself embodied the racial and gender-driven customs and traditions of the building trades. He was a plumber from New York City, and the son of a plumber who formerly headed New York City's Local 2 of the Plumbers' Union, a local infamously hostile to the induction of blacks and women.[78] Randolph tried to challenge Meany and the craft unions by broadening the focus of the black civil rights movement from strict identity politics to class-based issues. He successfully mobilized 5,000 blue-collar workers in 1963—both black and white—from all across the country to rally in New York City's Soho district to protest hiring practices and workplace conditions on job sites.[79] However, rumors circulated soon afterward in the AFL that the NALC and Randolph were dupes of the Communist Party, and Randolph felt compelled to disband the organization in order to, according to one of his associates, "maintain the correct posture with the AFL-CIO and Meany."[80] So, if much is owed to Van Arsdale's taking the lead among the building trades union on civil rights, the same (and more) can be said for the efforts of Randolph and the NALC. At both the national level and in the electrical trade in New York City, Randolph established a pathway for the generations of minority workers that followed.[81]

Ultimately, the attrition rate for the NALC's first class of black and Puerto Rican electricians in the brotherhood was high: only seven graduated to journeymen status.[82] Nonetheless, the minority fraternal clubs—the Latimer Society and the Iglesias Club—proved instrumental to the Local's labor-based civil rights political agenda.[83] Van Arsdale feared that growing racial strife over jobs might prompt governmental action to decertify the brotherhood's apprenticeship program and threaten the union's long-time control over labor supply. He was committed to bringing racial minorities into the trade, but only through the time-honored path of apprenticeship.[84]

Richard Nixon, the Philadelphia Plan, and the Brotherhood

The history of male racial minorities in Local 3 is incomplete without an understanding of the public policy pressures the electrical brotherhood and other building trades unions faced at the national level. Richard Nixon's election in 1968 shaped the civil rights struggle to integrate blacks into construction.[85] This story begins at the famous Archway in St. Louis, Missouri. In 1966, a group of white tradesmen walked off the building site when a black worker was placed on the job. The U.S. Department of Justice sued the building trades unions over the incident on behalf of the black worker, and as a consequence quotas and timetables for hiring black workers and other racial minorities became legal precedent.[86] The Nixon administration, seizing the opportunity to drive a wedge between the labor-based Democratic Party and its allies in the unions, sided with national civil rights leaders against the powerful building trades unions and established federal policies that required contractors to set goals and timetables for integrating racial minorities on federally funded construction jobs. This policy was called the "Philadelphia Plan"; in 1969 it became federal law.[87] The law required government contractors to meet minority employment goals for ironworkers, plumbers, steamfitters, electricians, sheetmetal workers, and elevator operators—trades in which minority participation rates were less than 1.6 percent.

The efforts of building trades unionists to influence the execution of the law in New York resulted in the adoption of a weaker "hometown" New York plan in 1970.[88] Developed as a result of negotiations among contractors, building trades unionists, and government representatives, the New York plan provided minorities with an alternative to the traditional craft union apprenticeship: minorities could be recruited as "trainees" rather than apprentices.[89] Trainees had to be paid at union apprentice scale, but building trades unions were not required to admit trainees to union membership.[90] Where it took six years to complete a regular apprenticeship, a trainee might need eleven years to become a journeyman, if it happened at all.[91] In the end, the electrical brotherhood and other building trades unions admitted very few trainees. Instead, Van Arsdale renewed his commitment to recruit male minorities into the traditional apprenticeship program. On October 5, 1971, he announced an affirmative action plan to be administered by the Joint Apprentice Committee (JAC) of the Joint Industry Board (JIB).[92] The plan committed the JAC to use the WDL, the Latimer Society, and the Iglesias Club to "enlist minority journeymen and apprentices to recruit applications from other racial minorities, and to interview them, and counsel them to enter and stay in our program."[93]

The brotherhood and the JIB soon found themselves engulfed in legal battles with the city, which forced their participation in the New York plan with a 1971 lawsuit.[94] As a result, the Local accepted one hundred trainees and agreed to make them eligible for journeyman status.[95] According to observers, these trainees were treated as second-class citizens and placed not in the elite Division A, but in Division M, where they became known among journeymen and apprentices as "M dogs."[96] A subsequent New York State Human Rights Commission complaint filed as a class action lawsuit on behalf of the trainees found in 1984 that the JIB had kept trainees separate from apprentices, gave them obsolete textbooks, and put them through a curriculum different from the electrical program.[97]

Despite success in court, city hall grew more impatient and frustrated with the ineffectiveness of the New York plan and the recalcitrance of most building trades unions.[98] In 1973, New York City Mayor John Lindsay issued Executive Order 70, which required craft unions to have a one-to-four minority ratio on all city-funded jobs and to increase their minority membership to 25 percent by 1976.[99] Another political and legal battle ensued: Local 3 and the Electrical Contractors Association (ECA) sued the mayor for exceeding his authority.[100] The brotherhood enlisted the support of Peter Brennan, former head of the New York Building Trades' Council and newly named by President Nixon as secretary of labor, to press Lindsay to relent. Brennan froze all federal funds earmarked for building construction in New York City, and the mayor relented.[101] In the end, the New York State Court of Appeals ruled in 1984 in favor of the electrical brotherhood, making Executive Order 70 illegal and reinstituting the New York plan.[102]

Social Clubs and Integration

Fraternal clubs in Local 3 have served to create ethnic and geographic "precincts" for monitoring electrical work, and knit together a diverse and temporary labor force of construction journeymen and apprentices.[103] The clubs generally reflect the immigrant succession and ethnic layering of New York City's construction workers, and various ethnic groups' organized attempts to create an occupational niche in the industry.[104] Scottish and German Masons dominated the Electrical Square Club of the early 1920s. To challenge the hegemony by Scots and Germans over industry building trades jobs, Jewish electricians organized the Electrical Welfare Club as a means of obtaining tradesmen's jobs and integrating into industry.[105] The Irish-dominated Catholic and Allied clubs and the Jewish Electrical Welfare Club began to gain influence and power in the industry and brotherhood in the 1930s, commencing with the consolidation of the Van Arsdale regime. The Santiago Iglesias club emerged during the late 1950s as a vehicle for communication for the brotherhood's electrical manufacturing workers who were predominantly Hispanic; while the Latimer Society started as a result of the 1960s civil rights era.[106] The Asian American Cultural Society was founded in the 1990s, followed more recently, by a fledging club representing newcomers from Asia.[107] Just as the various minorities have different histories within the union, their clubs have different relations and varying degrees of influence with the Local's power structure.

As discussed previously, whether the Latimer Society is actually effective in recruiting black men into apprenticeship is a matter of opinion. Some view the club as co-opted by the white male leadership. Others view it as useful. Claimed one black unionist:

The electricians are the most militant of all the trades, and the smartest. They understand the labor movement and how important unionism is. The Howard Latimer men push the union to bring more black men in but in a roundabout way. They aren't outwardly militant but more political. And they move like they are in the military, where chain of command is important. The club has influence.[108]

Whatever the general reputations of the different minority clubs, it remains true that the union used its established system of fraternalism to control the admission of black and Latino men in response to antidiscrimination legislation, although the union rationalized their induction into the brotherhood as just a matter of fair ethnic representation and not merely a matter of public relations.[109]

Over time, there has been a shifting of position for some of the clubs. The Latimer club still meets once a month but has relocated meetings from Queens to Harlem. Fightback still attracts the more militant black electrical tradesmen, while some Latimer members have become part of the union's internal operating structure, even though that structure affords them little meaningful power or influence.[110] Others have hardened their resistance to such assimilation: "The blacks in the union don't want to rub elbows with the institution of its agents. They are fearful and it's comforting to stay with your own, in the background, so that you don't stand out more than you already do," observed one black journeyman.[111]

The rift between the militants and other black electricians, like that between Fightback and the Latimer Society, is especially harmful now that organizations such as WDL and RTP no longer exist to recruit young black men from inner-city communities into skilled craft apprenticeships and jobs. Government enforcement of affirmative action has diminished and racial polarization has increased. Nonetheless, despite the factionalism, black men and other racial minorities have made greater gains in electrical construction than white women or women of color.[112] One African American female journeywoman sums up the race-driven challenges black women confront: "Here's the pecking order in the trade and it's pretty much true for all of construction—first is the white male, then the black male, especially when a woman is around, then, the white woman, and the black woman is last."[113]

On the whole, even critics such as Haughton agree that efforts to bring minority men into electrical construction in the 1960s and 1970s started well. Throughout this period the percentage of new minority male registrants in Local 3's apprenticeship classes averaged 17 percent (12 percent African American and 5 percent Hispanic), even though the recession of the 1970s precluded extensive recruitment of apprentices for all but three years (1974, 1978, and 1979).[114] Government enforcement helped as well. According to Haughton, "You had some semblance of this fighting spirit in affirmative action agents, in the Equal Employment Opportunity Commission

(EEOC), and they were actively pursuing affirmative action in the unions. There was some good government enforcement, and some government officials actually went down to talk to the unions and contractors."[115]

Title VII of the Civil Rights Act of 1964 proved a forceful instrument to break down barriers for blacks and other racial minorities, and eventually for the handful of women in the trade who followed on their heels. "I wrote a letter to Mayor Bloomberg just recently," Haughton says, "to activate New York City Local Ordinance 15 and State Ordinance 15a that states that there should be no exclusion on a work site on the basis of color." He goes on to recall, "In those early days, Major John Lindsay issued Executive Order 70, which had a great impact on the industry. It mandated that the contractors stop discriminating against men of color."[116] Others remember Executive Order 70 as short-lived with little impact on advancing men of color in the industry. In a 1993 Human Rights Commission Hearing in New York City, James Dooley, the former director of the New York plan for trainees, testified, "Out of the 5,000 trainees placed on jobs as tradesmen between 1971 and 1988, only 800 were accepted into the union."[117] That was disappointing, of course, but still an unmistakable boost to African American workers and an inroad for other male racial minorities. The story changed in the 1980s, illustrating the folly of relying on good intentions and assuming some "inevitability" of civil rights progress. After the first cadres of black and Hispanic men achieved their toehold in the trade during that remarkable window in the 1960s and 1970s, compliance reports show that their share as new apprentice registrants actually decreased during one of the most expansionary eras in skyscraper work in New York—the mid- to late 1980s.[118]

Predicting Apprenticeship Needs

Union contracts in the electrical construction industry usually run three years.[119] Overall, the size of the apprenticeship programs follows trends in construction activity. During negotiations, both sides try to predict the number of apprentices that will be needed during the term of the agreement.[120] The union errs on the side of caution, trying to restrict the numbers of new apprentices to protect journeymen from unemployment or wage cuts. But contractors, who want a large supply of apprentices since they are a cheaper skilled labor force than journeymen, would prefer they not err on the side of caution.[121]

According to data compiled by the New York State Department of Labor, the number of registered apprentices in construction trades in the New York/New Jersey region dropped by 60 percent between 1974 and 1976, and then grew steadily after 1978, the year following a twenty-five–year peak in the region's unemployment.[122] From 1978 to 1986, the number of apprentices grew by more than 150 percent. Between 1980 and 1986, the program completion rate for apprentices rose from 43 percent to 57 percent, resulting in 25 percent more graduates for a given cohort of apprentices. Suddenly, there were more workers than jobs.[123] And the Local's reaction to this oversupply of labor brought its preference for white male apprentices over minority men dramatically back to the surface.[124]

There is evidence of increasing discrimination against African Americans in construction during the 1980s, especially in higher skilled trades like electrical and carpentry work.[125] The elimination of Executive Order 70, the eventual demise of the New York plan, and a successful legal action by New York State's Department of Labor against five electrical contractors, all under contract with a small electrical union—Teamsters Local 363—for paying apprentice wages to trainees heralded a defeat for the forces of affirmative action in New York City.[126] Within Local 3, internal politics interfered with and slowed the pace of inclusion for black men. One factor contributing to this was the death of Harry Van Arsdale in 1986. The unofficial story and hearsay as reported by several insiders in the Local is that without his charismatic leadership, internal rivalries for trades work among electricians intensified along the divisions established by the club system: ethnicity, religion, nationality, family connections, and geographic locales. Families long established in the Local scrambled for what remained of the brotherhood's former power, and some took over particular clubs. For example, former Local union president Dennis McSpedon and his sons took control of the Westchester Mechanics;[127] while the McBurn family with ties to the Van Arsdale regime still controls the predominantly Irish Catholic Allied Club which dominates work in Queens.[128] Building trades work in midtown Manhattan was always an important locale for electricians to earn high-time money. This area, formerly under the control of the Electrical Square Club (originally a Masonic club dominated by men of German-Irish-Scottish extraction), is now under the control of the Jewish-dominated Electrical Welfare Club.[129] The Bedsole Club in Sheepshead Bay, predominantly Italian Americans, claims Brooklyn as its territory; the Acorn Club carved out territory in the Bronx; and the aforementioned, predominantly Irish, Westchester Club operates as a political club in Westchester County.[130]

Traditionally, membership in the clubs had not only been a prerequisite for promotion to business agent, foreman, straw boss, and steward, but also provided a venue to recruit young men into the trade.[131] Leadership as well as prospective journeymen in the industry are drawn from the clubs, and they serve as the functional arms for the Local.[132] As one African American journeyman observed, "whatever goes on beneath the surface of the union is carried out by the stewards. It's almost like a secret society."[133] This recent transformation of the clubs into vehicles for dynastic and factional struggles for political and economic power in the industry has left the Latimer Society with even less influence than before, and unable to compete for territory or effectively recruit blacks into the trade.[134]

As the civil rights movement weakened and organized crime infiltrated minority coalitions in construction, a new propensity for violence was introduced.[135] The new leadership tended to view minority men as a disruptive force on the work site and a potential nonunion workforce for contractors. Further, the inroads blacks made in the trade did not automatically result in an increased number of black contractors, which had been the trend among the Irish, Jews, Italians, Hispanics, and Koreans.[136] Since hiring in the construction industry is mostly by small contractors and subcontractors, and referrals are based on social contacts and who you know, the absence of black contractors as a critical mass inhibited the hiring of black workers.[137]

Finally, a paradoxical economic factor that developed in the 1970s—the oversupply of labor, which was a correction for an earlier undersupply—would disturb attempts to integrate minority men into the profession. The process began with the severe recession in 1976. Low wages in the late 1970s led many white working-class men who might have expected to become tradesmen to seek other local work options, while others relocated to other parts of the country and the world.[138] Jews, for example, and later on Italians, began to seek alternative work options and higher education.[139] They subsequently developed ethnic niches in such white-collar professional jobs as law, medicine, and civil service. Many journeymen traveled to other parts of the country during the recession, mainly to Texas, which was still in the midst of a building boom.[140] Some tradesmen traveled as far as the Middle East, especially to Saudi Arabia. Those tradesmen who remained in the city found that heightened ethnic and racial tensions on job sites became commonplace. As Randolph had predicted in 1964, white workers eventually blamed racial minorities and immigrants for deteriorating working conditions and increasing unemployment.[141] "I remember the 'good old days,'" recalled a black journeyman, "the severe isolation, the graffiti and racial slurs, it was very tough going."[142]

By 1980, many white journeymen were returning to the Local, in part due to the anticipated expansion of jobs. As the decade wore on and college tuition rose while federal student loans became more costly, apprenticeship, especially in the elite electricians' trade, began to attract more college-bound white males, many of them happy to take advantage of the generous tuition reimbursement provided by the Local as a benefit to members.[143] This trend was actually reinforced by the results of legal battles won by women's organizations such as All-Craft: the successful challenges to the use of age to reject women applicants opened the programs to male applicants who had some higher education and were older than the traditional apprentice age of 18 to 22.[144]

New waves of immigrant groups to the city also added to the labor supply of skilled tradesmen, and sometimes played to employer preferences. For example, Irish skilled tradesmen newly immigrated to New York in the 1980s had some advantages over other immigrant groups and racial minorities.[145] They spoke English and looked American. They also often had well-established ties to existing kinship and fraternal networks in construction.[146] This was especially true in the electrical brotherhood, where Irish-dominated clubs like the Allied welcomed fellow Irishmen as tradesmen over other immigrant groups and racial minorities. Subsequent waves of immigrants from Eastern Europe, especially Russia, brought in skilled workers already trained in their home country. By 1990, foreign-born workers made up 46 percent of the New York construction workforce.[147] Black electricians were crowded out by white immigrant newcomers and native white males claiming former prerogatives who were able to take advantage of continuing discriminatory practices.

Discrimination can be ameliorated when its victims mobilize and fight back. Unfortunately, the WDL, the main organization by which black construction workers mobilized politically and through which black applicants were recruited for unionized apprenticeships, became plagued by factionalism.[148] This limited the WDL's effectiveness in recruiting minority men into the trades.[149] It also crippled an important vehicle for countering the widespread resentment of what many white journeymen perceived as preferential treatment for black electricians. During good times and bad, contractors on public projects—especially federal projects—had to meet racial (and gender) quotas in hiring. As more people competed for the available jobs, affirmative action seemed more unfair than ever. One rationale offered by union leaders for the serious decline in minority admissions during the 1980s—that minority admissions were restricted out of fear that if minority electricians were hired and then laid off, the union and the contractor could be sued—may be plausible.[150]

There have been many racial discrimination cases filed against building trades unions and contractors since the late 1970s. For instance, in 1981 a class action suit was filed against Local 3, the ECA, and the New York State AFL-CIO, among others.[151] The plaintiffs (Samuel Lopez and Charles Calloway) alleged discrimination against black and Puerto Rican members of Local 3 by union officials in job placements and in grievance procedures, and named some thirty defendants including officials of state and federal agencies, various employers, and Local 3, charging a conspiracy to deprive plaintiffs of their constitutional rights in a variety of employment contexts.[152] The complaint alleged that the union intentionally discriminated against the plaintiffs by failing to process a grievance they had filed over a series of alleged wrongful terminations. The complaint also alleged that the employment assignments by the union had been discriminatory. In the end, the state court determined that Lopez had failed to show discriminatory treatment by employers and the union. Calloway's similar complaint had a similar fate: the union was exonerated on all charges in federal court.[153] Both cases, however, generated a great deal of negative publicity against the brotherhood in the media and industry.[154]

Another case involved Antonio Cancel, a Hispanic pioneer electrician and former member of the Iglesias Club, who had entered the Local in 1968.[155] In 1986, Cancel filed a complaint against the brotherhood for racial discrimination during a layoff. Testifying at a 1990 New York City Human Rights Commission Hearing about the manner in which the Local handled his case, Cancel stated: "My life was threatened by two men who put a gun in my face, and told me to drop the charges against Local 3."[156]

Thus, while the honorable imperative of integrating electrical construction work deteriorated from within as the legal infrastructure to enforce it withered from without, the brotherhood retreated from its leadership role in desegregating the industry. From 1990 to 2000, African American male labor participation rates in electrical construction stagnated at 10 percent (see Appendix B, Figure B.6), while newer immigrant groups such as Hispanics from Latin and Central America and Asians (mostly Chinese) increased their presence in the trade.[157] The return to white male hiring patterns and the immigration of new groups trumped the years of activism associated with the civil rights movements of the 1960s, and short-circuited the progress of inclusion for blacks.

This history is relevant to the integration of women, because women came into Local 3 on the coattails of black men.[158] The 1978 Executive Order that amended Executive Order 11246 mandated that female participation rates be counted apart from minority hiring rates in general on federally funded construction projects.[159] This order prompted Van Arsdale and Local 3 to voluntarily admit the first class of women into the apprentice program that same year. The future looked hopeful; by 1981, women comprised nearly 7 percent of the workforce on construction projects supported by the federal government—much higher than today's 2 percent.[160] That window of opportunity slammed shut once President Ronald Reagan took office. He appointed Clarence Thomas head of the EEOC, and Thomas immediately proceeded to remove from the commission's jurisdiction oversight of the rate of acceptance of minority men and women's applications into the various construction trades' apprenticeship programs.[161] He also gave oversight to the joint industry boards within each trade to monitor their own compliance.[162] With the teeth effectively knocked out of the antidiscrimination legislation, women in the electrical trade and the industry generally thus never even approached half the 13.8 percent penetration rate currently reported for minority males.[163]

Emerging Trends: Women and Minority Men

Despite working women's interest in better paying jobs; the perseverance of the brave women pioneers; the campaigns by advocacy organizations; and executive, legislative, and judicial intervention, women remain largely excluded from the construction industry today. Other industries and professions have become gender-integrated in the past few decades, but not the building trades, which remain 98 percent male (see Appendix C, Table C.1).

Some women prefer electrical work because it seems to them to be more intellectual than other trades. They also envision electrical work as a jumping off point for a career in engineering or drafting. But male unionists and contractors argue that women are simply not interested in this line of work and that it is a forced fit—something unnatural that government is trying to enforce. Contractors also argue that women are discouraged from the industry because of the harassment they experience from men on the job—a situation they view as inevitable and impossible to change.

The fact that minority men have been more successful than their female counterparts in getting into the trades, though they continue to be underrepresented, has been thoroughly discussed in this chapter.[164] Minority men pose no threat to the gender symmetry of the construction brotherhoods, the manliness of the trades, or a male-dominated domestic gender hierarchy. While racism remains endemic in the trades, minority men, according to one white male electrician, "have more of a right than women to be in the industry."[165]

Minority male electricians report that much of the racial harassment on the job comes from workers from other trades, though the harassment of blacks and Latinos in the electrical industry is certainly documented.[166] It is possible that as their numbers grow in the trade and industry, Hispanic men will face increasing hostility.[167] Yet the harassment of women far exceeds the nature, scope, and intensity of these racial and ethnic incidents. To begin with, union officials and contractors alike view racial and gender harassment differently. Unionists and contractors will denounce racial harassment on the job, but they accept the harassment of women as a natural consequence of the presence of females on construction sites. "If women want to put themselves out on the line like that, then they have to take what's coming," says one contractor.[168] Since contractors rarely provide a nonhostile work environment, women are often afraid to go to work. One journeywoman declares:

> There is no guaranteed protection or center for the women to get help when they need it right away. Absolutely nothing exists. Women should have immediate attention when they are harassed on the job sites. That does not exist in any of the construction trades. It should be set up like the rape hotlines or the suicide hotlines and it's not. Women should run hotlines for harassment so that they can be made to feel safe where they work. Building companies should be made to fund these.[169]

Legal challenges have sometimes exposed collusion between contractors and union officials to keep women out of the industry. In 1991, Annette

Streeter, a woman of mixed Hispanic and Native American heritage, and Ivette Ellis, a Hispanic woman, sued Local 3, the JIB and the ECA for sexual discrimination.[170] Streeter and Ellis had been terminated from an electrician's apprenticeship program in early 1988. They alleged that at several of their work placements they had been subjected to verbal and physical sexual harassment and other discriminatory treatment, including the lack of separate changing facilities. They also alleged that they were discriminatorily terminated by the electrical contractors on several of these work sites, and then discriminatorily terminated from the apprenticeship program of the Local. Streeter and Ellis won their suit; moreover, the court held that the JIB, the Local, the Joint Apprenticeship Training Program, and the ECA were all together liable as an "integrated enterprise."[171]

Compared to many of the lawsuits against the Local for racial discrimination, gender discrimination seems much easier to prove in court, making legal blockades an unlikely explanation for the lack of success for women's integration. The different trajectories of women and minority men are more likely to be found in culture and custom than in legislation and jurisprudence. In comparison to women, minority men challenge very few of the ideological assertions of the electrical brotherhood. Minority men are family breadwinners, deserving family wages. They can meet the physical demands of the job and endure the dangerous working conditions. Indeed, while unionists and employers may disagree on many issues like wages and job classifications, they can agree to view women as a sexual diversion and hindrance to the efficiency of electrical building work.

And so, within the structured fraternal system of the union and the male occupational culture of electrical construction work, women are anomalous. This very anomaly threatens the fraternalism that knits together a highly contingent, culturally diverse, and geographically dispersed labor force, and which historically supported the discipline and solidarity that gave the union its power. On a more practical level, women entering the elite construction division of the brotherhood pose a significant challenge to its political hierarchy. If one combines the number of women in the manufacturing division and the elite construction division together, women outnumber men in Local 3; if they voted as a block they could oust any current leadership.

Race and Gender on the Job

If minority men have gained some grudging acceptance by their white counterparts, women have encountered a ruthless and persistent culture of resistance. Once grudgingly admitted to Local 3's construction division, women have faced obstacles not placed in the way of others seeking a place in the Local's ranks. Add in governmental indifference to enforcing affirmative action or contract compliance, and the struggle to integrate women into the trades faces yet further impediments not present when the task was integrating minorities. Male electricians explain this in different ways. For example, women are thought to have higher attrition rates in the industry.[172] Though not proven, this assumption suggests that letting women into the apprenticeship program wastes precious opportunities.[173] Male electricians also complain that women have gained entrance through "outside interference," threatening their sense of union autonomy.[174]

Of course, minority men came in the same way, but they are now considered "ordinary hard-working guys," and how they got into the union is easily forgotten. One white male unionist puts it this way:

> When you talk about blacks or Hispanics or any other male coming in, you know they come in to do the work. It is synonymous with the craft and it is synonymous with the manual work in the trade. But what we see happening when women come in is that they enter not because they are looking for this type of work but because they are steered in. It is a nontraditional job, and here is an opportunity to take a whack at it. Some of them came in mainly to test the system, and when they are accepted they just drop to the wayside. They are not interested.[175]

Solidarity across racial and ethnic lines not only helps the union prevail in conflicts with employers, but also helps work crews complete difficult construction tasks where mutual support and trust are critical. The need for this bonding seems to have somewhat trumped the racial resentment common elsewhere in society. "In an all-male group, there is a lot of 'father-son' type bonding," a white journeyman explains.[176] White electricians have apparently learned to accept men of different races into their "work family."

Then "women come in, and sexuality just fucks it all up."[177] White men can approve, even admire, minority men adapting to the macho culture of construction, accepting danger and poor working conditions without complaint. For many minority as well as white men, the high wages and status of the craft more than compensate for such discomforts. But the first generation of journeywomen (many of whom had middle-class backgrounds) proved much more likely to see contractors as responsible for health and safety standards—and were more willing to rock the boat over these issues.[178] Perhaps men relished the dirt and the danger as validation of their masculinity and justification for their wages. Perhaps they were embarrassed to reap the benefits of women agitating for better conditions for all. They might simply have been uncomfortable considering women in a new light. Working-class culture often styles women in contradictory ways: on one hand, special, pure, and rarefied; on the other, sexy, sullied, and besmirched.[179] Men may share discomfort about women performing "dirty work" across racial lines but the unease springs from different sources. For white men, male supremacy over the women in their lives and breadwinner status in the family motivate their disdain for women in this work. Black men may well share these family values, but they also worry that black women may become targets for sexual attacks or disrespect by white males.[180]

One African American journeywoman laments that black men "are the hardest ones to accept you on the job. . . . I can get everyone's respect; I can even get a Latin male's respect before I can get my own black brother's respect." It seems clear to black women that black men's resistance to "their" women on the job has more to do with assertions of masculinity, and acknowledging the fraternal bonds achieved with white workers in a racist society, than with competition with women for overtime or better jobs.[181] Black men may also fear that black women will besmirch their hard-won status as skilled workers. They may worry that women's lack of experience at physically demanding work will reflect poorly on all blacks and risk costing them their own jobs.[182] One African American woman, in the electricians' trade for over 20 years, thinks so: "The white guy wants to get the women laid off, so they make a pact with black men, and the black men go along with it thinking they are part of the clique. But little do they know that after they get the women get laid off, they [black men] are probably next. Black women talk all the time about how the black men are used to knife us in the back."[183] Such dynamic gender solidarity seems to require a betrayal of racial solidarity, though gender solidarity may not pay off in the end for minority men.

Nonetheless, macho attitudes generally prevail over racial solidarity between black men and black women. "You will run into those who will side up with the white males just to be part of the bunch," reports one African American journeywoman. "The whites' attitudes are the same toward them [black men], except now they have this one thing in common." Another concurs: "There are a lot of cases where I feel like the men would be fighting among each other but when it comes to an issue about a woman, then all of a sudden there is unity against the woman."[184] Cross-racial male gender solidarity also may be encouraged by the reluctance of minority tradesmen to risk being perceived as troublemakers. Local 3 officials recognized the activist politics of ethnic and racial organizations as potential threats and tried to divert that focus by encouraging minority members to join union-sanctioned social clubs and groups, thus splitting the ranks of blacks and other minority men.[185]

Perhaps minority men in Local 3 exploit the presence of women on the job to cement their solidarity with white men because solidarity within their own ranks is in a delicate state. As mentioned earlier, African American and Hispanic (and now Asian) male electricians have long sought to establish unity by forming grassroots organizations that help promote them into craft trade work.[186] In the late 1970s, at the height of their integration, women comprised about 2.3 percent of the Division A electrical apprentices. This low percentage continued through the 1990s. Compared with other craft occupations like carpentry, participation rates for women have actually fallen since 1978; at the same time, minority men have risen from less than 5 percent to more than 20 percent of the field. In electrical construction, minority male participation rates in electrical apprenticeship programs went from 5 percent in 1978 (the first year Local 3 accepted female apprentices) to 15 percent in 2000.[187] Despite persistent cultural, racial, and ethnic rivalries, men of color and diverse backgrounds have been able to transcend, at least superficially, the barriers of race and ethnicity and bond together on the job, and as brothers within the Local. Whether the presence of women on the job facilitates this male bonding is an open question.

The appearance of any woman in the industry still appears to infringe on male privilege. Of the thirteen thousand powerful and prestigious Division A electricians, approximately 1,800 are minority men (apprentices and journeymen) and approximately 300 are women (both white and minority).[188] While initially minority men were perceived as a political and economic threat to the journeyman's racial privilege and especially his ability to pass along his craft to his sons, women continue to pose an even greater challenge to the customs and traditions of the electrical construction brotherhood. It is time to acknowledge that what has been rationalized as the customs and traditions of fraternity and brotherhood have been used to hold back the advancement of women in the union and the industry.

6

A Club of Her Own

I was talking to one of the guys on the job who is active in the clubs about some of the problems we have in the Women's club. He turned around to me and offered some advice. He said, "Ya know, sometimes you go for the homerun, but sometimes you've got to step back and punt, and that's just the way it is. You just can't expect to hit home runs all the time." And he said, "You just can't expect to go out there, make a demand and get it."[1]

Women presented a problem to the leadership of Local 3 from the beginning. Journeywoman Evan Ruderman, from the apprentice class of 1978, recalls that when she and three other female pioneers first entered Division A, business manager Harry Van Arsdale Jr. made a speech to members of Local 3 regarding the need for safe and cheap public transportation.[2] He spoke about the crowded subways of the time and said that workers do not want to be crushed against each other. Then he added, slyly, "However, I'm sure the women don't mind that too much."[3] It sent the overwhelmingly male union hall into gales of laughter. Union members began boisterously chanting the business manager's name in happy approval of his words. The four female members walked out, shouting protests. They demanded a written apology from the manager. According to Ruderman, "We told him that we really didn't believe that he felt that way in his heart and we tried to give him a way out." They received no response. Then, four months later, the manager told another conclave, "Fellas, the girls here think that we are not giving them the recognition and respect that they deserve. I want each and every one of you to treat them fairly on the job."[4] Nothing changed. The words of even the most powerful man in the union could not modify the attitudes that prevented union men from accepting their sisters as coworkers and colleagues.

The women apprentices could see how their white and African American male colleagues banded together in clubs and organizations.

These groups offered tangible assistance with work and school, helped the apprentices gain respect and to become full partners with union men, and provided a social and community link to the larger world of work.[5]

There had long been women's auxiliary groups designed to keep women, in the old Masonic parlance, "outside the door."[6] The first Local 3 group organized by and for female workers was the Women's Active Association (WAA), formed in 1954 for the women in the union's manufacturing division.[7] Thirty years later, Martha Midgett, a former official of the still-extant group, recalled its original aim: "to fight for social equality and encourage political action on behalf of women's rights."[8] She praised the union leadership's "progressive" role in making its formation possible and boasted, "We existed long before many of the women's rights groups."[9] However, she also recognized an important element in its constitution: the WAA always affirmed and supported Local 3's leadership and initiatives.[10] "We always believed," she said, "that unity is necessary to create and maintain an effective union . . . and have consistently supported the officers of Local 3. . . . They have done all within their power to address issues brought to their attention."[11]

The union leadership agrees with that interpretation. An article in the Local's official *Electrical Union World* calls the WAA "a forum where women members . . . come together for the purpose of identifying working conditions in the shops that are unacceptable."[12] Like the reform-minded labor feminists of an earlier era, the women of the WAA felt that issues of class and unity took precedence over issues of gender equality. Most of these women leaders in the union come from the divisions other than construction like electrical manufacturing and supply shops in the boroughs of Brooklyn, Queens, and the Bronx. Most are black Americans primarily from the Caribbean, reflecting the majority of women union members.[13]

When the women apprentices in the construction division first entered the union in 1978, they looked into joining the WAA but reportedly discovered that the association's main activities were shopping sprees. In fact, WAA leaders rarely raised issues that affected women working in the manufacturing divisions; their loyalty was to the Local. Women from other divisions did not challenge the Local's view of gender stereotypes. Today, the WAA appears to exist in name only. Past leaders have retired and not been replaced. Most women now in the skilled construction division of Local 3 have never even heard of the group.

Yet the new women electrical apprentices wanted to be accepted into the trade and the brotherhood on an equal footing with the men, and to distinguish themselves from women workers in the manufacturing and other lesser skilled divisions of the Local. In 1978 they formed the women's club of Division A called Women Electricians (WE) (often commonly called the Women's Club), the only club in the Local that ever attempted to remain outside the union's control. Though now disbanded, the club's original leaders and some of its members are still active organizers of tradeswomen in New York City and nationally. Former club leaders joined other tradeswomen in 2000 to create a new organization, Operation Punch List (OPL), affiliated with the feminist organization Legal Momentum (formerly, the Legal Defense and Education Fund of the National Organization for Women) and comprised of 250 tradeswomen across various building trades such as electrical work, carpentry, pipefitting, steamfitters, and plumbers to name a few.[14]

Meeting weekly, OPL helps women "shape up" for construction jobs, both union and nonunion, and provides an educational and support network for women across a broad spectrum of building trades.[15] OPL was instrumental in the founding of a national tradeswomen's organization, Tradeswomen Now and Tomorrow (TNT), which held its founding convention in New York City in 2004.

The history and evolution of WE within the Local is a microcosm of women's struggle to form bonds of sisterhood in a brotherhood based on the fraternal customs and traditions of the previous century. The club served as a study group for apprentice courses and as a support group for women who were experiencing all-too-typical isolation and harassment on the job site. During the 1980s, most newly indentured women joined its ranks. From its beginning, WE challenged long-standing practices of the brotherhood and in return got nothing but trouble from the leadership of Local 3.[16]

That first class of educated, politically committed, feminist apprentices were determined that their club would not duplicate the way the other major fraternal clubs in the union were run, nor be controlled by the central administration of Local 3. That control supported the union's traditional strategy of stratifying new entrants in the trade into hierarchies. The pioneer women wanted to challenge that hierarchy of inequality in society generally, and particularly in the union and the industry. Leaders of WE, while remaining committed to the labor movement, voiced constructive criticism to the leadership and attempted to engage union women in "outside" feminist causes. But their primary agenda focused on challenging inequalities and work conditions on construction sites, and pressuring the Local to recruit and retain a greater number of female apprentices. These activists had deep reservations about becoming too involved or too close to the Local's power structure for fear, in the words of one journeywoman, of "being separated from our sisters."[17] This was one of the most notable differences between the first women members of Local 3 and those that followed in their footsteps. The pioneer women remained more aware of class, gender, and racial privilege in the brotherhood and the industry than younger women. Complaints from the early female members about younger women in the trade focused on their flirtatious ways with men on the job which, in their view, "give the women a bad rep."[18]

Many of the issues the club's members considered important were issues taken up by the broader feminist movement: pornography, sexual harassment, and gender equity issues generally. Such concerns were quite alien to the union culture of manly solidarity. The Women Electricians club further antagonized the union by insisting that the club would be a safe space where issues and complaints of all kinds, whether against the contractor or the union, could be discussed without fear of retaliation.[19] Pioneer Laura Kelber, whose father held a staff position at the Local as education director, remembers: "We were willing to take a risk and put ourselves in a situation that was totally hostile to us. We knew that it was more than just our own personal interest—it was bigger. We felt we had a right as women to do this. We were taking a stand."[20]

The Women Electricians club was perceived by some union members as too radical in nature to serve as a unifying force for all the women who worked construction in Local 3. It did, however, serve as the very stage upon which the most crucial questions were played out about what female integration into the construction trades should mean, and how it should be accomplished. Its factionalism, its eventual demise (in 1999), and its replacement by a sanctioned club of female electricians with kinship ties to the union and the industry, all attest to the continued powerlessness of women working in male-dominated jobs.

The Women Electricians club was not especially attractive to many of the women who entered construction work seeking strictly economic security for themselves and their families. They were primarily concerned with escaping an economic market where remuneration was stratified by gender and race. They disliked the idea of a radical organization aiming to reform the union that they had worked so hard to join, a union in which most of them had relatives, and through which they had been given the chance to learn a skilled trade. Poor women felt more loyal to the Local's leadership and black women wanted to be less conspicuous among a sea of white males on the job; both groups just wanted to fit in and were suspicious of the intentions of the women leaders of the club. As it became evident that male union leaders were indifferent or hostile to feminist concerns, it seemed safer for union women to watch the feminists than to join them.

In 1987, the club successfully negotiated with the leadership to hold a conference for the women members of Division A. Its organizers recall that some male union leaders assumed that the conference would be a chance for women to "trade recipes" or discuss "the soaring price of meat." Instead, the women issued a list of desirable reforms and recommended a labor education program and campaign to better inform contractors, foremen, and rank-and-file workers of women's legal rights to work in the trade.[21] The union failed to follow up, demonstrating the Joint Industry Board's (JIB) lack of interest in educating union members about gender integration in the trade.

A few months after the conference, a group of female apprentices confronted apprenticeship director "Buddy" Jackson. They asked that the union start addressing sexual harassment and general male hostility to their presence on the job. Jackson replied, "You're pioneers in this industry. You're dealing with this job out there individually, and that's about the best that you can do." He might as well have added, "And we're not here to help you."[22]

Most union members shared the attitude that "if you want to be in a man's business, you'd better accept our rules and culture."[23] The Women Electricians club activists—even those fighting for broad-based reforms that would help all construction workers—were not treated well when it came to important matters like job assignments. Union officials made it clear that they preferred women who joined the quiescent, dormant, but officially sanctioned WAA instead of Women Electricians.[24]

The introduction of feminism into the construction world spurred a backlash among blue-collar electricians and their leaders. Men asserted and cultivated their fraternal bonds. Shifting definitions of gender roles sparked a negative reaction on the part of male leaders. They viewed women's presence in the electrical division as alarming—personally unsettling and politically divisive. "The reactions of men we see sometimes on the work sites are due today to changing definitions of what it is to be a man, and what it means to be a woman.[25] Nobody knows today, it's all mixed up," lamented a former president of Local 3.[26] The men's clubs became increasingly popular. Men openly disparaged their female colleagues.[27] One high-ranking union leader inveighed against women in construction: "It comes about as some kind of contest. I mean, they're not there primarily because they have a tremendous desire to be there. Traditionally, they have not been exactly adaptable or suitable."[28]

A former JIB official explained that while union leaders tolerate existing women's groups in the manufacturing division,

> construction is the heart of the union. . . . Division A women will not [be allowed to] get into leadership positions. And the reasons for this are cultural, political, and psychological. Some male union leaders [in the building trades] say, "Women don't have the time, they are not attuned to leadership, you don't want a woman with two kids because you know that there will be an emergency meeting and she will say she has to be home with the kids."[29]

Tradeswomen would counter that the real reason male union leaders fear female union leaders—especially women from feminist organizations—is because they are less likely to have family connections in the trade and therefore less likely to be loyal to the union. Such women could not be disciplined by solidarity-conscious brothers, uncles, or fathers when they criticized the union or the unionized contractors. Evan Ruderman, a member of the second class of women electricians, who was present at the meeting with the JIB apprenticeship director, observed, "The men that were part of the 'union family' felt that they had a great deal of investment in the union. Their grandfathers had started it and fought for its present structure, and now they are part of that structure."[30]

Cynthia Long, the first Asian American woman in the Local's construction division and a member of that first class of women pioneers, put it a little differently: women are a threat because they "can come in and say that this is not a democratic organization. Whereas on the other hand, if you grew up in the Local, you realize that it was your bread and butter from the day you were born. You have a different sense of right and wrong. They are really afraid that any change will bring about destruction. They fear that women will take over their positions."[31]

The union's fraternal organizations have always acted as effective mechanisms for control. They provide a setting where dissidents can be identified. They provide access to positions within the union and to more favorable assignments on the job. "The union defines someone as 'active'—a highly desirable designation—on the basis of whether they belong to the clubs," according to Ruderman. "The clubs are also ways to obtain overtime, become a foreman, or work your way up the ladder."[32] By encouraging community and welfare activities, the clubs limit the possibility of widespread criticism of the leadership.[33] The Women Electricians club was different. It rejected the rigid rules with which the union hierarchy limited the fraternal clubs and provided a forum for critical discussion of the union and its leadership. As a result it was seen as a threat.

At first, Women Electricians gatherings served mostly as gripe sessions—a great boon to the growing numbers of women apprentices who needed an outlet for their on-the-job frustrations.[34] These gatherings echoed the larger feminist movement's "consciousness-raising" sessions. The women found that their experiences were shared by others and that their problems were social, not individual. They learned that they had not caused the problems they faced. This freedom to discuss women's conditions in the industry and to articulate women's social needs and contributions was central to the Women Electricians club. Melinda Hernandez recalled: "Its purpose was really for emotional and moral support and to share information so that each of us could achieve journey-level status. It was formed to help each other survive—that is, help ourselves by helping each other because every woman who leaves creates a huge vacuum."[35]

Sometimes Women Electricians members had to make some noise. Long remembered: "At union meetings, when some union official would say, 'Okay, you guys, this is what we're going to do,' we would say, 'and women.' And the men would say, 'Okay, you guys, you should bring your wives to such and such a function,' and we would say, 'and husbands and boyfriends.'

In other words, we would constantly remind them that the women were there."[36] But it hardly followed that the women wished to destroy the union or to overturn its leadership. In reality they wanted to strengthen the union. "We wanted to stay and make the union better," Laura Kelber recalled, "more responsive to the problems of all construction workers."[37]

This, of course, is consistent with the "second wave" feminism of activists like Betty Friedan, who asserted that fulfilled women make better parents and better partners and that feminism helps everyone, not just women.[38] Some journeymen grasped this point. Kelber continued: "But we perceived early on that the men, although sympathetic to the issues we raised, did not support us because they were afraid of losing their jobs. They may want to criticize the union, but they don't do it. They generally don't feel it's worth the sacrifice to change things the way the union works."[39]

An Unsanctioned Entity

As the number of women apprentices and journey workers increased in the mid-1980s, the Women Electricians club became an unsanctioned caucus in Local 3. In contrast, pioneer black male electricians never formed an unsanctioned group like the women did when they entered the Local, except for NALC, and then later Fightback, both of which operated outside the union structure.

Meeting monthly in the homes of its members, Women Electricians provided a safe space for women with differing identities—African American and European ethnics, working class and middle class, lesbian and heterosexual. Eventually the club split along lines of race and sexual preference, and many women stopped coming to the meetings.[40]

African Americans were the first to leave. They joined the Latimer Society and attempted to build a coalition with black male members at the IBEW level.[41] "The Women Electricians club should have addressed the issue of race," commented a veteran Latimer Society journeyman. He continued: "The blind terror that women face in an all-male environment like construction traumatizes them, and black women don't want to stand out more than they do already."[42] But some remaining members of Women Electricians felt that the black women's departure was not motivated by only racial issues: "among this early group, we were all stereotyped as gay by men for being in the trade."[43]

Because the men felt that these women were "out of place," they also thought that they "must be gay." In the construction trades generally in New York, there is a significant division among women between "out" activist lesbians and heterosexual women, a division that transcends racial boundaries.[44] Some heterosexual women view lesbians as just another clique trying to promote their own agenda. What is perceived as cliquish may, however, be a coping strategy of lesbian tradeswomen. One journeywoman suggested that "the gay women are completely unacceptable to the male membership but they tend to make it through in larger numbers than any other group. Gay women hang out together, and lend support to each other; they tend to play down other differences."[45]

Nonetheless, in the electrical trade as well as the other building trades such as carpentry, the male leadership uses differences among women to retain their own power and privilege in the union. Both real and perceived divisions between lesbians and heterosexual women in the building trades tend to stand in the way of women's progress to recruit and expand female membership.

In the electrical trade, Women Electricians members differed in how they related to the Local along lines of race and sexual orientation. For example, African American women wanted to associate more with the men, white or black, and to get involved in sanctioned union activities such as social outings, community work, walking picket lines, or surveying and reporting on nonunion jobs in the boroughs. They feared that their club association, with its unsanctioned status and its affiliation with outside feminist groups, would be a mark against them in the union. As one black journeywoman and Latimer Society member observed, "direct action worked in the 1960s and 1970s, and created a stink to get women and blacks into the trade; but within large entrenched organizations today, you're just going to say to the leadership that you are a marked person."[46]

Nevertheless, the club's unsanctioned status had real advantages. Progressive feminists in the club like Ruderman remembered how she and her colleagues "did things as first-year apprentices that guys would never dream of doing, such as petitioning the union, and calling meetings with the apprenticeship director. This is what kept the women bonding."[47] For these activists, the union's sanction had some disadvantages. Hours of volunteer work were required of members of sanctioned clubs, while women workers were apt to be already committed to family and child care, and had less free time. Cynthia Long, a founding member of the club, recalled:

> I observed from the sanctioned clubs in the Local that the union leadership feels free to call upon you any time of the day or night to turn your members out to do whatever they want you to do—for example, show up at a demonstration, stuff envelopes, things like that. But I felt that there was no way I could have such control over these women's lives, and I didn't want to have to guarantee 20 women on the picket line.[48]

"Besides," she later added, "I wasn't interested in doing things without knowing why. And that was their thing. They would say, 'Show up at such and such a place,' and they wouldn't tell you why. They [the Local] would say, 'jump' and you would have to say, 'how high?' "[49]

If progressive feminists, black women activists, and lesbians shared a common vision for workplace gender equality, they worked for job site reforms in different ways. African American women, who were more likely to come from working-class family backgrounds than progressive white feminists, expressed some empathy with the men. They were interested in constructive action, not what they perceived as "carping."[50] Dorothy Mays, an African American journeywoman with family in the trade, remembered:

> I got involved with the Women Electricians in my first year. I was looking for women who saw what I saw, and I didn't find that. I did not find any of the same attitudes. I felt like an outcast, an outsider, because I was looking at things from different points and needed to say to them, "Have you tried different strategies with the men?" For example, a woman comes in and we all tell our problems, [which are related to] shanties and bathrooms. The men used to drill holes in the wall. But what the women failed to realize was that most of the women who got the shanty in the first place had gone down to the union hall and to the shop steward. So the guys were retaliating. At the meetings you would have eight girls having the same stories and it was hard for me not to say to them, "Hey, ease up." I thought their attitudes were wrong. I used to say to them, "Hey, we are women, let's approach things the way we have always approached them with a civil attitude, and really looking at the situation before we dive into it and sink."[51]

Another African American journeywoman, who served on the advisory apprentice committee, recalled: "Many of us were tired of hearing the same talk about harassment and pornography—and they are still talking about these things."[52] She thought that fighting to get more women to be shop stewards, foremen, and union representatives on negotiating committees would be more useful than campaigning against sexual harassment or pornography.[53]

Many African American women electricians suspected that white women, especially feminists, were wrong to assume that the confrontational tactics of the 1960s and 1970s feminist movement would be effective in breaking the systemic union culture of exclusion. As the journeywoman above put it, "If you want these things to change, you have to come out and be active in the union." Shirley Merriman-Patton, a member of the Latimer Club, drew a sharp contrast between the "inside" and active nature of her club and the unsanctioned status of the Women Electricians club:

> I was on the advisory apprentice committee and we were going to functions, learning things, going to the Waldorf. I was out there dealing with the NAACP and the Negro College Fund. The fact that

the Women's Club was outside the union was a problem for several of the black women. Every time you turned around, the white women were going to a demonstration or to a coalition meeting. But when we would ask them if they were going on the apprentice picnic or on a ski trip, you know, things that were with the guys so you don't cut yourself off from them, they would always say no.[54]

Black women did not want any more problems than they had already. "They wanted to socialize, and they wanted action and a way to solve problems. We felt that we should find out what was going on in the union before going outside the organization. We felt that the white women were using tactics that large organizations use, while the union is a very personal place." Progressive feminists were looked on as arrogant: "They see themselves as 'trendsetters and 'in a class by themselves.' "[55]

Mary Au, a Chinese American electrician from an immigrant family who came into the apprenticeship program through the municipal trainee program in 1986, shared the sentiments of African American women about the Women Electricians club. Progressive feminists, it seemed to her, had little exposure to working-class culture. Consequently, "their expectations about advancement in the union are simply unrealistic." Before becoming an electrician, Mary had tabulated truckloads for the Teamsters, so she had some perspective on sexual harassment. "There are all types of harassment, not only sexual," she noted. "For instance, not being able to get the proper training during the apprenticeship was more harassment for me than dirty pictures on the wall."[56]

Despite the rifts in Women Electricians, the members' common experiences as women gave them a special sense of closeness. They shared work stories, strategies for survival, and harassment tales without fear of retaliation.[57] The club fostered the development of a woman's work culture by sponsoring picnics, beach outings, dinners, Christmas parties, and baby showers for tradeswomen and their families. As journeywoman Ruderman recalled, "We were attempting to build a sisterhood that could connect women, not just as electrical workers, but, in the broader sense, women struggling for equality in society."[58]

This struggle was important to them. Many Women Electricians club members thought that to gain union sanction and a place in the brotherhood's hierarchy, they would have to renounce their feminist focus as well as their freedom of expression and association. If the club were sanctioned, it was feared that women electricians invited to speak outside the union as representatives of the Women Electricians club would have to obtain permission from the union and clear the content of their remarks with union headquarters. The progressive feminist club founders viewed this as too restrictive and were unwilling to relinquish their autonomy. Eventually, this decision by the club's leaders to remain independent from the fraternal structure of the brotherhood left them with few options to promote gender equality in the trade; instead, they had to rely on outside organizations such as governmental regulatory agencies to enforce compliance or to sue for the redress of grievances.[59]

The Male Reaction

Male union activists tended to view any attempt to resolve "private union matters" in the public arena as anti-union behavior, and the decision of Women Electricians to remain independent outraged male unionists. Ruderman recalled: "The union dealt with criticism of any kind by arming itself for battle. . . . People who were active in the union really feared the destruction of the union."[60] Outside agencies could only promote division and create weakness.

Moreover, feminism was not just any outside force. Male labor leaders viewed feminist organizations with great suspicion and women's issues as "special interests" that would only divide the membership.[61] They insisted that women downplay their gender identity and prove themselves simply as workers—no different from men.[62] "I would say that there are many women who are physically able, and personality-wise, they're tough," suggested a high-ranking union leader who spoke on condition of anonymity. "But the thing is—are those the women who will be seeking job opportunities in the construction industry? Or will it be some little girl who really isn't suited to this big-job situation?"[63] Naturally men like this were hostile to the Women Electricians club and warned off potential female apprentices, especially the daughters, sisters, and nieces of journeymen, who began to enter the union in greater numbers in the 1980s and 1990s.

More to the point, from the Local's perspective, Women Electricians was unsanctioned rank-and-file organizing, an activity disloyal to the union. How reasonable was this concern? According to club leaders, a critical mass of women in Division A could place women on an equal footing with men and transform the union's internal structure.[64] Thus the handful of women activists "swimming against the tide?" in Division A's sea of male electricians posed a threat to the male unionists who then held power. Women Electricians could also potentially galvanize women in the other, predominantly female, divisions to aspire to leadership positions in the brotherhood. The privileged white male inheritance—father to son to grandson—of a good living and a respectable place in society could become a thing of the past if women, especially minority women, won union leadership. Also, if the number of women reached critical mass in the construction divisions then the union would be forced to raise issues in bargaining with unionized contractors that could escalate their costs: parental leave, pregnancy disability, sick days, better sanitation, and a general upgrade of job site safety and health conditions.

When women attained journeyworker status and became one of the "elite," men resented them even more. One working-class Irish American journeywoman remembered that by the time she became a journeywoman, she had been on the job long enough to believe she had weathered the worst abuse that male workers could dish out.[65] She found, however, that she was "going through the prejudice all over again with the guys; as an apprentice I wasn't so much of a threat, but because now I make the money they make, they don't like it."[66] Women faced even more hostility when they aspired to "run work"; that is, get a foreman's job or obtain a position as shop steward or run for an elected union office.

In an effort to gain respect, the club's leadership held a second Division A Women's Conference at Bayberry Land in Southampton in 1990 and a third in 1992.[67] During these gatherings, conference organizers wanted to bring *all* women union members together, whether they were affiliated with sanctioned men's groups and clubs or not, or whether they considered themselves activists or dissidents.[68] By getting Local leaders and JIB members to attend, the organizers also expected to give female apprentices and journeywoman greater access to the top male leadership of the union and the industry so they could directly address problems on work sites. However, after the 1992 conference the Local decided to create a parallel women's club, the Women's Committee of Division A, an initiative supported by at least one of the pioneer women for the purpose of addressing problems women electricians faced on work sites. Many feminist unionists regarded its creation as an attempt to dismantle the more militant Women Electricians club. The fact that the Local reserved the right to appoint Women's Committee officers seemed to confirm the women's capitulation to the male power structure.[69]

Struggle for Democracy

The unity of union men across ethnic and class lines is most obstructed by the go-along-to-get-along ethic of the union itself. Consider an incident from the mid-1980s involving Melinda Hernandez, a feminist Hispanic from a working-class background, and three other Women Electricians club members.[70] The four tradeswomen were working on a prestigious job site in the city when a black woman traveler was physically attacked by a white male coworker in a shanty. Hernandez rallied the women on the site to complain to the foreman and the shop steward. She was warned by her male straw boss that she would be removed from the job if she pursued her complaints. Initially supported by the other women, Hernandez then wrote a letter requesting the removal of the offending man. As a result, both she and the black woman who had been attacked were removed from the job. The other three were able to remain in return for dropping the issue, thereby reneging on a promise to write in support of Hernandez. One of the three eventually became a foreman and then a union delegate.[71]

Women in the construction trades have found it very difficult to address their persistent mistreatment by male coworkers. Without their own political base in the union, tradeswomen have relied either on appeals to union leaders or the expression of their indignation at public hearings to redress their grievances. Neither course has had much effect. Indeed, women electricians credit persistent male hostility as one of the reasons that so many women leave the trade prematurely, or are discouraged from entering.

Women have had to depend on the enforcement of federal legislation and court rulings to make any gains, exactly the kind of outside interference so resented by the brotherhood. This reaction is somewhat illogical, since the brotherhood already depends on government interference for a number of benefits and economic subsidies. For instance, the Davis-Bacon Act guarantees a union wage for workers on federally funded construction sites, and the National Apprenticeship Act of 1937 (also called the Fitzgerald Act) mandates and partially funds the skilled apprenticeship programs and the union monopoly on the labor supply. In the economic crisis of 2008–2009, electricians and other building trades in New York City and nationwide rely on President Obama's economic stimulus package to create needed infrastructural building construction jobs and job training programs. Nevertheless, government hiring quotas for women and minorities on federally funded jobs are viewed by nonunion and union contractors as an unwelcome outside interference. This not only impacts differential access to good-paying jobs for women in general and women of color but there also exists a correlation between the number of women in these field jobs, women's access to higher-level positions in construction management, and their ability to establish successful small-business contracting firms in the industry.

By the 1990s, those still active in the Women Electricians viewed women's presence in the trade as an issue of workplace fairness and gender equity. Pornography and sexual harassment remained important issues, but the tradeswomen also wanted to improve health and safety for all construction workers. For example, on federally funded job sites such as the pollution and sewage treatment plants where women and minority workers are likely to be assigned, women electricians voiced concerns about health and safety to contractors and the Local.[72] Women Electricians members also petitioned the JIB to address such issues as separate changing and toilet facilities and reproductive hazards on the job.[73] They neither triumphed in these initiatives, nor were they successful in getting the union to address family issues such as parental leave.

There are too few women workers on construction sites to foster mentoring relationships like those found among the men. Without an alternate base of solidarity such as a unified, sanctioned, and effective club, the female apprentice can find herself isolated and ignored. Playing along with the guys, ignoring issues of safety and health, not rocking the boat—this is a plausible strategy for survival and advancement. This was the experience of women who entered in the trade in the 1990s and beyond, and an occupational hazard of the male-dominated workplace. It did not promote sisterly bonds. A white journeywoman who entered the electrical program in 1996 explained,

> When you're on a big construction site and you are the only woman, for good or bad, you make a niche for yourself. Then, when another woman comes on, you immediately feel threatened. Which is ludicrous, because there are two women out of, like, 1,000 male workers? But once they accept you, you're like their mascot, and with another woman there, you're not going to get as much attention.[74]

Even if they understand women's grievances, male electricians may not support them out of the fear of losing their own jobs, or at least their advantageous job assignments. Word gets around about so-called troublemakers who file complaints with the Local. A veteran of thirty-one years in the Local and the only remaining journeywoman of the first class of women who entered Division A in 1978, Cynthia Long observed:

> There is a tremendous boys' club underground. This does not necessarily come from the leadership, but it exists among the workers, especially the foremen. Phone calls are made before you arrive to a job, and this is true for both male and female workers. Information in this underground is disseminated about workers prior to their appearance at job sites in an attempt to single out troublemakers.[75]

In 1979, however, during a period of severe economic uncertainty in the industry, male apprentices did unite with women in a risky reform effort when many apprentices—black and white, male and female—supported Long for an officer's position on the apprenticeship advisory council.[76] According to Ruderman, then an apprentice, both male and female apprentices were up in arms over the announcement that the last year of the apprenticeship requirement was to be extended from twelve to eighteen months. She recalled:

> The men tried to figure out how to organize against it. A petition was written, but the most dramatic display of opposition came at a union meeting where the extension was announced to an auditorium of nearly 2,000 apprentices. After the announcement, the guys got really angry. They began to yell and scream. They stormed the auditorium and ripped the doors off the hinges. There was literally a riot.[77]

The women apprentices supported their male colleagues because it was in their own interest to do so, and because many felt that apprentices might finally be able to act in concert as a result of this action by the Local. The incident touched off a movement for apprentice voting rights and a contractual agreement on apprentice employment. Ruderman recalled, "The whole incident was a really good example of how indignant and outraged apprentices felt at the whole process and the lack of democracy," the very things the Women Electricians club was targeting for reform.[78]

However, the movement for apprentice voting rights came to nothing, in part because the apprentices were not able to enlist any journeymen to their cause. Long, Hernandez, Kelber, Ruderman, and other pioneer apprentices at that time remember that the journeymen argued against apprentices having the right to vote because, in the journeymen's words, "they would try to take over the Local." They felt apprentices did not know enough about the rewards of journeymanship to appreciate the union. Hierarchy won out, but for the first time men and women had worked together within the union for a common cause. This fleeting attempt at unity illustrated the fragility of class relations in the brotherhood: given the right circumstances, the underlying class interests of apprentices and journeyworkers could trump ethnic and gender divisions.

In 1988, at a divisional journeymen's meeting, Laura Kelber, now a journeywoman, attempted to nominate Cynthia Long to be a district representative. One eyewitness recalled that at first the president refused to recognize Kelber's hand, but the men in the audience started shouting and he was forced to do so.[79] She nominated Long for the position, and it was seconded by the men. Without the guys, Long would not have been nominated at all. After the meeting, many male apprentices approached her to request that, should she be elected, she raise important issues with the leadership. Long agreed. By the next meeting, the underground had done its work, and many workers refused to pick up her flyers describing her platform. Nonetheless, she maintained a large measure of support from minority men. When the vote was tallied, Long lost by fewer than one hundred votes—the first time that a candidate not picked by the leadership had come close to winning an election.[80] Long's cross-gender support suggested a slowly emerging solidarity among male and female white apprentices as well as a willingness by minority men to support women of any race who would champion their issues.

Women continued to raise important social issues during the 1989 contract negotiations, albeit with limited success. Women Electricians sent a letter to the negotiating committee requesting that it include issues of parental leave and sexual harassment in the next negotiation.[81] Each woman mechanic was asked to bring the letter to her Division A meeting and ask her male coworkers to sign it. Long described the leadership's reaction to this tactic:

I was talking to this guy, who was a shop steward, and he wanted to sign the letter we drafted, but then he became uncertain and didn't want to sign it. I told him that I knew how he felt about these issues and that he should take a stand on them. We were standing up in the back of the union meeting and talking about this. From the stage, I saw the president whisper to somebody and then signal them. They went to the security. When they finally came around to me, they told me I wasn't allowed to do that in a union meeting and that it wasn't the Women's Club meeting so I had to put the letter away immediately or I would be removed from the hall.[82]

Many of Division A's progressive feminists believed they should advocate on behalf of low-wage women in the Local's manufacturing, telephone operator, and lamp and shade workers divisions.[83] The theory was that building unity with women in the manufacturing divisions, which are overwhelmingly female and therefore more likely to support women's issues, would ultimately strengthen the position of Local 3 tradeswomen overall.. "As women workers, I'm sure we have a lot of things in common," Ruderman argued. "So I don't really feel it's helpful for us to buy into this thing that a lot of men buy into, where we're a different division, and we're the highest division, and our problems are so different and so unique or whatever. I think it's a kind of arrogance. I would like to see us have more contact with other women."[84]

To this day, the Local's leadership fears that Division A women might organize across the Local's divisions using women's issues. That would subvert the union's hierarchical structure, putting the predominantly female manufacturing divisions on a footing equal to the now-dominant construction division. In Hernandez's words,

It would make the union aware that they are not just dealing with a few loud-mouthed women from the construction division. They're dealing with a broader issue that has to do with women in the workplace. I think it would be helpful for women in manufacturing to understand the working conditions of women in construction, as well as it would be for us to understand theirs. People would probably be surprised, and then they would probably be disgusted, at some of the conditions that we work under. And I'm sure they would give us their support to change it.[85]

According to Local 3 leadership, Division A has the most power because it generates the most funds for the union. However, the other divisions, composed mainly of minority workers, comprise the vast majority of Local 3's membership. One journeywoman electrician who entered the Local in 1981, related, "I have always been of the opinion that, deep down somewhere, the local union leadership fears that if these women get organized, they could easily threaten their power."[86] Indeed, many of the men perceived Women Electricians as powerful enough, in the words of one male journeyman respondent, "to organize all the women against us."[87]

The Local leadership also feared that Women Electricians would work with outside organizations and agencies beyond the federal bureaus, such as the state Human Rights Commission and the Legal Defense Fund of the National Organization for Women.[88] In fact, such outside alliances gave the Women Electricians club what little power it actually had through their ability to lobby for better conditions for women workers. But even this power was strictly limited. For example, in 1990 in New York City, tradeswomen acting in coalition with the New York City–based feminist organization Legal Momentum and minority advocacy groups such as Fightback and the United Bridge lobbied the New York City Human Rights Commission to hold public hearings on race and gender discrimination in the building trades.[89] The hearings were organized through the Women's Project of the Association of Union Democracy (AUD) and aimed at publicizing the exclusionary recruiting, hiring, and training practices for women and minorities in the skilled building trades, as well as the contractors and unions' failure to address these conditions.[90] The hearings exposed contractors with discriminatory practices, and pressured public officials to withhold federal- and city-funded building contracts until joint industry and apprenticeship boards complied with federal, state, and local ordinances for race and gender goals and timetables for inducting women and minorities into apprenticeship programs.[91] AUD also hoped to influence city hall to develop a municipal hiring hall that would curtail what they perceived as the "monopolistic" control over the construction labor supply, as well as the corrupt "featherbedding" practices commonly found on many construction sites. None of these reforms have come about.[92]

Many of the progressive feminists who entered the trade in the late 1970s and 1980s were inspired by earlier movements for equality. But for a majority of women who followed in their footsteps, direct action was a strategy of the past; they viewed compromise as the only way to gain ground on the job site and in the union. Even among some of the progressive feminists from that first class, political action as opposed to the direct action strategies of the 1960s and 1970s seemed a more effective way to "bring more men to their side" and shift the balance of power. They did not deny that their role might always remain a disruptive one in union councils, which they considered a good thing. Since working in construction is itself for a woman the result of successful struggle, women workers tend to be bolder than men about raising important issues: "Women come from a different place than men, so we're not afraid to take risks. Women take more risks than men anyway, just because being a woman; you have to if you want to get what you want out of life. They take more risks in relationships; they take more risks in everything."[93]

Frustrated by failing to achieve change either with contractors or with male unionists, Women Electricians disbanded in 1999. Club leaders and some club members formed Operation Punch List (OPL), an advocacy organization comprised of tradeswomen across a broad spectrum of skilled building trades. Currently affiliated with Legal Momentum (LM), which is the former Legal Defense and Education Fund of the National Organization for Women, OPL meets once per month in LM's downtown Manhattan headquarters.[94] OPL is a networking resource for women interested in entering unionized apprenticeships, or "shaping up" (going to work sites as day laborers) for construction jobs or other nontraditional blue-collar work, both union and nonunion. It also provides support for tradeswomen who are taking legal action against employers and unionists and holds conferences to increase and retain the number of women in the trades.

The next generation of women electricians would try a different strategy to recruit and sustain women in the industry. These women tried to assimilate into existing customs and traditions of the brotherhood. Did these alternative strategies work, and would this next generation of women electricians achieve greater success in integrating women into the prestigious construction Division A?

The Next Generation: The Amber Light Society

Suspicious of Women Electricians's links to outside organizations, Local 3, as previously mentioned, started a second women electricians' club, named the Women's Committee of Division A, aimed primarily at women electricians with family in the industry. The year the Women Electricians club disbanded, the Women's Committee became the Amber Light Society (ALS), a sanctioned women electricians' club, the female equivalent of the Latimer Society. Regardless of its position within the fraternal structure of the Local, its members report that the society has virtually no power.[95] ALS leaders have indicated that unlike the officers of other clubs they are not included in regular meetings with union leaders, union contractors, and master mechanics held at union headquarters to discuss issues in the union and the industry. Furthermore, unlike other clubs, including the minority clubs, ALS does not serve as a training ground for potential leadership in the Local.

ALS operates mostly as a social and philanthropic club.[96] The Local's male leadership occasionally attends ALS-sponsored social events, but there has not been any real progress on addressing women's concerns: harassment on job sites, denial of training by journeymen, attrition, recruitment of women into apprenticeship, and promotion to leadership positions. According to a former ALS leader, the club's size inhibits its effectiveness: "There are hundreds of people in the other, older clubs," she observes, "but only 40 women in Amber Light."[97] Nonetheless, the society's mission, as described by another member, is to:

> address the needs of women in construction, many of whom are single mothers, and women need a forum to address problems that they have both on and off the job. . . . We meet once per month, and the women in Division A bring their issues to get resolved. We also opened the club to men and encourage male union leaders to address the group. Usually, the leadership will show up if it's necessary to dispel any rumors about incidents involving women on job sites. We've enjoyed some support from the Local and the men. And over the years, we have become more involved in the community and with charitable organizations like the one we work with in Long Island and Queens, which is an organization for children whose mothers are in prison. But it's clear we are not part of any decision making, and we do not operate like the men's clubs—sharing information about work and jobs or moving up in the union. We're supposed to bring our issues and our problems to the club for the Local to solve. But we've never gotten them addressed, and we're really doing mostly community work these days.[98]

There are some parallels between ALS and the minority male fraternal clubs in Local 3. Indeed, an analogy can be made between the more independent and militant Women Electricians and Fightback, on the one hand, and ALS and the Lewis Howard Latimer Society, on the other. But the differences are profound. ALS did not become sanctioned until women had been in Division A for over twenty years; the Local legitimized the induction of black and minority men by immediately creating the Latimer and Iglesias clubs. Moreover, male unionists still tend to view ALS as a ladies' auxiliary rather than a fully functioning fraternal club, and not without some reason. ALS performs philanthropic work while minority male clubs focus on industry-wide problems, the growth of nonunion work, employer challenges to work rules, and changing relations between journeymen and apprentices that may affect work assignments.

For white male electricians, the minority clubs have given their members some legitimacy in the brotherhood, but these same journeymen view ALS the same way they saw the Women Electricians club, as "a bunch of dykes getting together."[99] According to one longtime member of the club, "the Amber Light Society is supposed to offer women in the community counseling services, big sister programs, and models itself"; in the words of another ALS member, "after Mother Teresa."[100] ALS members, however, have begun to challenge this image and role as a ladies' auxiliary. In a recent meeting, ALS members discussed the treatment of women on the job and the need to instruct women entering apprenticeship about their rights on the job and in the union.[101] Issues such as discrimination, harassment, pregnancy leave, and lesbian partner benefits have begun to appear on the club's agenda. In the 2006 contract negotiations with the Electrical Contractors Association (ECA), the Amber Light leadership successfully fought for birth control to be included in health care coverage—which benefits men too—though they failed to win maternity benefits or light duty for pregnant electricians. Paradoxically, even though ALS members do not necessarily subscribe to the ideology of "sisterhood," lesbian and heterosexual women work together in the society, and recently club members supported a lesbian member's petition to the union for partner benefits in the industry.

Amber Light Society members view themselves as a "bridge for women electricians in Division A" between the Local and contractors. "Contractors are on the Joint Industry Board in the same building with the Local," notes an ALS member, "and many current contractors were former Local 3 apprentices and journeymen, so as you can see, the relationship is very intimate." When it comes to women's complaints about work in the field, she continues, "a construction desk for complaints for men in the field already exists, [so] ALS helps reconcile differences between men and women in the industry."[102]

Parallel to the Local's use of the minority male clubs, Amber Light members are invited to attend the apprenticeship interviews at the JIB. "We refer questions that the women have in the Division to the men in the Local, and there is always a woman at the interviews for new apprentices," states the first woman foreman in the Local.[103] The club does work with one women's organization, Nontraditional Employment for Women, which which helps to identify new recruits. As one member admits, the Local "could always use a little push recruiting and retaining women." Other ALS members are more critical: "We lose women all the time because they do not know who to call and what to do when confronted with adverse circumstances on the job." Club leaders have urged union officials to conduct an exit poll of women leaving the industry to identify problems and increase retention—so far without success. There is no special orientation for women entering apprenticeship, though Amber Light members "would like to speak directly to women apprentices when they first come into the Local."

There are now two female foremen in Local 3, but many members of ALS complain that although women have been in the trade for nearly thirty years, there are still no female shop stewards. Shop stewards could act as role models for younger women and understand the needs of women in the field; instead, the women can only "air their complaints" at ALS club meetings and gingerly point out the Local's "blind spots" when it comes to Division A female electricians. "Women have so many issues," summarizes one

ALS member. "We got together in ALS to have some solidarity. But we were even derided by the men about what we named the club. They said that amber is a derivative of copper in electrical work, and that copper is used to 'feed' electricity. They always have a way of twisting things to feel superior."[104]

In 2008 the society celebrated its eighth year as a female fraternal club within the brotherhood. Despite their kinship, both real and fictive, with men in the industry, ALS members feel their problems in the Local go unheard and that women remain outside the walls. Even though the club is sanctioned by the Local, its members find it hard to raise issues of gender equality and still appear as loyal unionists.

The brotherhood remains uneasy about women's presence and fearful of legal challenges. The Local now holds meetings of women electricians to air complaints, but this process only began in 2004. In 2007, two hundred women electricians were called to the JIB for one such meeting organized by Christopher Erikson, Harry Van Arsdale's grandson and the newly elected business manager. One journeywoman who attended the meeting recalled: "The union was up to its typical tricks. None of the microphones were working at the meeting so no one could hear each other or what the women were saying from the floor."[105] Nonetheless, women challenged the leadership about ignoring their issues. They complained about inadequate access to lockers and toilet facilities, and contractors' discriminatory practice of laying off women first, regardless of seniority.

A journeywoman and member of ALS describes another such meeting: "At the Bronx Courthouse last year all the women were laid off and replaced by men. A woman electrician wrote the Local to complain and the Local responded by holding a meeting of women electricians because they were afraid of a lawsuit."[106] At this meeting women electricians pressed the Local to conduct a survey of layoffs by gender, and to address contractors' violations of seniority protections. According to reports, union leaders professed ignorance of the violations and ducked the issue: "All the business manager could say was that he was 'new'."[107] For another journeywoman who has family in the trade and is a member of ALS, "the Local is just becoming more and more blatant—they don't want women in the industry and they are apparently not afraid to be more open about it."[108] When she asked the business manager if he would help women in the construction division with their problems on the job, assist in reducing the attrition rate for women, and help them recruit more women into the industry, "he turned to me and said, 'Tell me how many women are there in the construction division? And then he answered his own question: 'there are 300, I believe,' meaning that I shouldn't be bothering him with this small group. There weren't enough women in the division from his point of view to worry about."[109]

Members of the Amber Light Society are developing new strategies. At the last IBEW convention, some activist members reached out to the IBEW's women's and minority caucuses to form a broader coalition with Division A female electricians. The coalition's goals include getting grievances addressed, resolving problems on job sites, and creating leadership opportunities for racial minorities and women in the construction division.

So it appears that the Local's attempts to meet regulatory mandates while maintaining the traditional hierarchy may not have completely succeeded in maintaining control over women in the craft. Despite relying on family and related channels for recruitment, and despite creating and nurturing a sanctioned fraternal club for women electricians, the story of women in the electrical brotherhood is still unfolding and the next chapter in their struggle for equality has yet to be written.

Conclusion: Getting Women Down to the Job Site

S ince the early 1960s there has been a substantial increase in the number of women in the labor force, but it has not resulted in a more integrated labor force. Although women comprise 48.1 percent of the labor force as of 2007, more than 80 percent of all employed women work in just 71 of 400 detailed occupations.[1]

Examining women's entry into electrical work and its affiliated craft union brotherhood afforded a unique opportunity to gain insights into the ways in which transformed relations of fraternalism operate in the present day, particularly with regard to the role it plays in influencing hiring patterns, labor market composition, and the reproduction of occupational sex segregation. The formal and informal work and organizational structure of the electrical trade—such as fraternal clubs; the labor process itself on construction sites; the collusive interests of male contractors, unionists, and workers—operate to successfully keep women out, despite legal challenges and legislative reform. Nonetheless, the nonunion construction sector does not do any better by women. Associations and their lobby groups like the American Builders and Contractors (ABC) and temporary labor agencies such as Labor Ready undermine prevailing wage laws, discriminate against women, and jeopardize women workers' safety.

Construction union federations in some areas of the country like California are beginning to address the integration and recruitment of women into skilled trade apprenticeships. Although this is a hopeful sign especially since unionization substantially increases wages and job security for women workers, New York and the Northeast in general appear to be the most intransigent.

The significance of women's inability to make any noteworthy inroads into electrical construction also reflects, to some extent, the strengths and weaknesses of the social movements of the 1960s and 1970s in regard to equality, their precipitous decline in the 1980s, the divisions among feminists, and women's overall progress in the workplace.

Race also structures the dance between men and women in the male-dominated workplace. By drawing comparisons with the integration of other racial and ethnic groups into the electrical brotherhood at various periods of its history, the central theme of male solidarity emerges in multiple contexts to explain how privilege at work—based on class, gender, and race—reveals institutional and subtle patterns of sex segregation.

The integration of male minorities into the electrical trade also presents a paradox: Given the prevailing racist climate in the country as a whole, why have African American and other minority men been able to make inroads into the electrical brotherhood and achieve relative acceptance in the industry and on work sites, while women have not? Furthermore, why did the electrical brotherhood, while downplaying and resisting women's entry, show political leadership in the integration of minority men?

Institutional and work site customs and traditions of male solidarity among journeymen and apprentices, and their corollary paternalistic view of women, operate beneath the surface of workers' collective conscious to exclude women workers as "brothers" in the union. Male occupational culture poses a formidable barrier to women's inclusion. Case studies in such public sector work as firefighting and policing, and in private sector occupations such as mining, have documented in general the problems associated with masculine resistance to women on the "shop floor."[2] These case studies illustrate that hostile male occupational culture functions to the exclusion of women, and to a lesser extent, men of color. This case study of electricians has expanded on these prior works and raised additional questions: Why are men in the construction industry unwilling to tolerate changes in their work culture or relinquish any control over their labor organizations? How does the resistance to the presence of women on the job site manifest on the work site and relate to gender relations at home, and to what degree do the conditions of work shape gender relations on the work site?

This chapter contains a final look at the interplay of such factors as historical contingency, patterns of male solidarity, and occupational work culture in electrical construction in accounting for women's place and acceptance.

Job Desegregation by Sex: Workplace
Justice in Slow Motion

Efforts at leveling the playing field for women in the construction trades—and to enforce the Equal Pay Act of 1963 and Title VII of the 1964 Civil Rights Act—have moved at a snail's pace.[3] Despite an increase in litigation since the 1970s that continues to the present day, neither unionized nor nonunion employers have made serious efforts to eradicate discriminatory workplace practices in good-paying fields like the construction trades.[4] Contractors act in contradiction to the principal theorem in a labor market economy, which predicts that employers will eradicate inefficient systems and workplace practices because of their unwillingness to pay the high costs associated with such systems and practices, in this instance, discrimination.[5]

The absence of women workers on construction sites (except in a few rare instances) is emblematic of how so little has been achieved regarding women's inclusion in the trades despite so much legislative reform and advocacy. Sex segregation in the blue-collar world encourages conventional views of women as unsuited for such male-typed occupations and professions as engineering, drafting, and science, which require technical know-how and training in mathematics.

Sex segregation and stereotyping in occupations have commanded the focus of social scientists from across a broad spectrum of disciplines, including anthropology.[6] Human capital theory illustrates that differential labor market outcomes for men and women are found on the supply, not the demand, side. It espouses the notion that job segregation by sex is reflective of larger societal patterns of socialization that ultimately lead to differential valuing of women and men in labor markets.[7] Further, supply-side theories suggest that women need to invest more in training and educational preparation for nontraditional jobs like the building trades.[8]

On the demand side, labor segmentation theory presents a more analytical approach to explain sex- and race-segregated labor markets. For example, factors such as employers' "divide and conquer" strategies among male/female workers and ethnically diverse workers determine employer hiring patterns. In addition, recent accounts of women in municipal service jobs like fire service and policing indicate that pernicious male occupational culture is central to the reproduction of sex-segregated occupations.[9] Until now, however, no study of a contemporary occupation and workplace has attempted to explain how factors such as craft traditions of fraternalism shape the speed, or slowness, with which women's integration into the construction trades occurs.

Men, Masculinity, and Unionism:
Some Observations

Sociologist Ruth Milkman lays male labor organizations' stereotypes of women in nontraditional jobs at the feet of employers' customs and traditions in hiring. The unwillingness of male union leaders to break ranks with men over women in nontraditional occupations is common.[10] Like male unionists in the auto industry, most electrical brotherhood unionists fear that their advocacy on behalf of women workers will invoke the retribution of male membership and contractors. Milkman points out that one irate male member of the United Automobile Workers (UAW) union told the women in a Ford plant to "go home and tend to [your] knitting."[11] And, at a past union educational seminar, a male union contractor suggested that women electricians might want to "trade recipes."

In the electrical industry of the 1940s, only the United Electrical, Radio and Machines Workers of America (UE) promoted the view that the male union leadership's willingness to demote women in skilled jobs meant playing into the hands of a "divide and conquer" strategy of employers. A 1944 UE report relays a warning to male unionists that anti-union forces would use gender "to split labor's strength by creating distrust and dissension between men and women workers . . . and disrupt the solidarity of the labor movement."[12] The present-day gender war in the building trades has proven this prophecy true.[13]

Male workers, employers, and unions are resistant to women's inclusion in the skilled trades, and prospects for change are uncertain. In addition to the building trades, a number of other private-sector occupations, such as in auto manufacturing, mining, telecommunications, and the computer industry, remain sex segregated.[14] Recent court rulings have opened up skilled craft union apprenticeships to women and minorities especially in the auto industry but most success stories of union activism on behalf of women workers still emerge primarily from public sector unions such as the American Federation of State and County Municipal Employees.[15]

Although white women, as well as women and men of color, work for prevailing wages in skilled building trades jobs, women (and minorities) continue to be scapegoats for white male tradesmen's angry and confused responses to the nonunion movement in the twenty-first century. Both groups of workers have been blamed, historically, for the "corruption" of the craft system and for hastening the demise of traditional forms of building trades craft apprenticeship.

The presence of women, especially women of color, is often viewed by male colleagues as "contaminating." Some employers claim that women's sexuality will be a disruptive presence on building sites, slowing productivity, while tradesmen and unionists react to women as an unsettling presence in an industry and occupation that faces unprecedented challenges and changes due to globalization, such as the growth of nonunion contractors and overseas unionized electrical manufacturing, the threat of what already established tradesmen perceive as encroaching nonunion immigrant labor, and the shifting boundaries of jurisdictions among the various building trades.[16]

Gender, Race, and Fraternalism

The power of the electrical brotherhood's organizational strength reproduces gendered values and paternalistic concepts of brotherhood among Division A electricians. These values reflect deeply subsumed notions of a woman's place at work and at home.[17]

The traditional "place" of women in the brotherhood's manufacturing divisions is one illustration of this paternalism. Women workers are organized into divisions and jurisdictions which are much less politically influential in the union and industry. Yet women make up the vast majority of the electrical industry and of Local 3 as a whole. But in the male-dominated construction division, a journeyman, in addition to imparting the skills of the electrical trade, is also responsible for inculcating paternalistic "values" of "unselfish" brotherhood to a younger generation of male apprentices toward lower-wage workers, mostly women, in the union. Young male apprentices, white or black, who are sponsored into the electrical craft apprenticeship are imbued with an ethos of manliness and citizenship that places women in a subordinate role. As one male unionist, now an electrical contractor, stated: "Women are good for the telephone division [in the Local], because their smaller fingers can handle smaller wires."[18]

After World War II, the brotherhood felt compelled to organize women of color and people of different nationalities—on a paternalistic basis—in factories and electrical supply houses in outlying working-class communities: Queens, Long Island City, Brooklyn, and the Bronx.[19] This paternalism both hindered and benefited women manufacturing workers.[20] Women's separate sphere in the union did not always indicate weakness and victimization; at times, the union has been a great resource for women of color in the Local divisions aside from construction. Union loyalty in the 1940s and the 1950s forged identities among women factory workers that were just as strong as ethnic or neighborhood ties. Although they did not want to be defined as feminists, women in the manufacturing divisions developed a class-based consciousness regarding the treatment they deserved at the hands of employers.

Like reform feminists of a bygone era, described by labor historian Dorothy Sue Cobble, who were instrumental as unionists over the first half of the twentieth century, women of color in the less-powerful manufacturing divisions of the union worked to elect the status quo leadership in the brotherhood. Like the feminists of the 1970s whose values clashed with these reform feminists in the electrical manufacturing division, pioneer women electricians stressed women's individual and civil rights over strict adherence to union discipline. Accordingly, these pioneers challenged male unionists' sense of women's place and the essence of the Local's organizational governance structure. Although women electricians entered a predominantly female union in the electrical brotherhood, class and race divisions between women electricians and their sisters in manufacturing prevented them from developing a base of support.

Generally, present-day views diminish the importance of traditional gender relations, but not in the electrical brotherhood. Journeymen caution apprentices on the danger of women taking the place of deserving men with families to support. In addition, women are viewed as cheapening the value of the trade to contractors, thereby possibly lowering standards and wages. Construction Division A journeymen are charged with the responsibility of teaching their young male associates the tricks of the trade, which seem to include passing on values about the patriarchal family and the primacy of the male breadwinner's role in the family. A woman working on the job site simply reminds a tradesman that his economic superiority vis-à-vis other blue-collar workers may be at risk.

Traditional notions of male gender ideology lace every aspect of cradle to grave unionism in the electrical brotherhood, and the fraternal clubs in Local 3 also shape notions of gender relations in the electrical trade.[21] Incessant social rituals of these fraternities and the use of male-typed iconography such as banners, logos, depictions of knighthood, Boy Scout symbols, and paintings of heroic male saints such as St. George defeating the dragon, symbolize nostalgia for a preindustrial craft past that revered the work of craftsmen. They also help to reinforce the cultural mores of masculine identity and privilege and provide a reference for union solidarity.[22] Furthermore, the "play" aspect of fraternal clubs in the brotherhood (e.g., the sports competitions, the dinner dances, and the communion breakfast ceremonies) serve to bond male workers together outside the realm of narrow definitions of unionism.

The electricians' family housing and a workers' community of shops and markets known as Electchester were planned and built by the labor of Division A men working within their clubs. The workers' educational retreat, Bayberry Land, and the summer camps for electricians and their families, Camp Integrity, continue to teach the same union values to the next generation of workers including a gender ideology of the male breadwinner as head of the household.

On the job site, brotherly "chit chat" about club activities help to break up otherwise monotonous routines. When a worker is laid off, his loyalty to the brotherhood is secured by the many roles he plays as a union member and team player.

The fraternal clubs, divided along the lines of ethnicity, race, gender, geographic location, and occupation, also serve a practical purpose; they provide vigilance over the encroaching nonunion sector and serve as proponents for volunteer activism of its members. Both functions facilitate the central administration of the union by, in the case of the former, guarding the jurisdiction of electricians, and in the case of the latter, providing more cost-effective service to the membership.

Local 3 is not the only building trades union or industry with fraternal clubs, but it has by far the most highly evolved system of clubs.[23] Although these fraternities can be found throughout the thirty divisions of the brotherhood, the Division A clubs are the most powerful and influential in the Local and throughout the industry.

Clubs are also influential in shaping recruitment and deployment of journeymen and apprentices in the industry. Informal kinship networks and apprenticeship referrals in the trade have been the pattern for most of the Local's history. As ethnic layering of immigrant groups provided opportunities for earlier European immigrants to obtain professional jobs in occupational niches in the city, father to son sponsorship gave way to include nonkinsmen and workers from diverse racial and ethnic groups. Mother to son sponsorship into the construction division by women of color in manufacturing permitted minority men to gain a toehold into the coveted electrical apprenticeship A program. As the civil rights movement gained momentum in the 1960s, the union identified alternative forms of male sponsorship from such organizations as the National Association for the Advancement of Colored People (NAACP). This pattern of pluralism was aimed at reducing possible ethnic conflicts on job sites and thereby saving contractors the expense of time and money to discipline workers.

In place of clearly demarcated exclusionary policies, typical of nineteenth- and twentieth-century fraternal associations in America, the electrical brotherhood manages its version of inclusion by placing members into ranked fraternal niches: European American men occupy the highest-ranking niche and men of color and women are at the bottom. Ethnicity and race play a pivotal role in constructing these niches. Researcher David Horowitz defines *ethnicity* not merely as an element of that person, but a relation between groups in specific contexts.[24] Horowitz categorizes the construction of identity based on ethnic factors into what he terms ranked and unranked systems. In ranked systems, groups stand in clear relations of super- and subordination to one another, positions that are often linked to particular occupational positions. He further suggests that ranked systems are more resistant to change than unranked.

The ranked hierarchy of racial and ethnic clubs of Local 3 determines their access to information and political power in the trade. Tradesmen's acceptance of their subordinate positions within the ranked fraternal structure of the brotherhood in combination with the seasonality of the industry and the temporary nature of workers' relationship to employers drive workers toward unquestionable loyalty to the union. Given these conditions, the union and the JIB provide Division A electricians with a sense of community and continuity, and a potpourri of benefits, that make workers fiercely loyal to a profoundly unequal and exclusionary institution.[25] The stakes are high: outstanding benefits (especially in comparison with other building trades), prestigious work, and top pay contribute to the fierce competition for these good unionized jobs.

The electrical brotherhood invents "traditions" of unity based on customary tradesmen practices of male gender solidarity that knit together ethnically and geographically diverse tradesmen in the late nineteenth and early part of the twentieth century. Today, male solidarity is still very relevant but takes a much broader form and application. It not only safeguards the interests of workers to protect standards and their employability, but it also applies to unionists who wish to maintain a caste-like hierarchy in the local union. Male solidarity in the electrical industry continues to function in the interests of employers who wish immediate access to highly skilled journeymen and lower-paid yet skilled apprentices needed to fulfill goals and timetables for the completion of large and complex construction projects. The 1960s social movement for black male equality almost short-circuited this caste-like hierarchy, but the electrical brotherhood and the joint board weathered the storm of racial discontent only to sail a decade later into the choppy seas of women's liberation and equal rights.

The 1960s and 1970s Struggle for Equality

In the electrical industry, the civil rights movement had an important effect on the inclusion of minority men—and later on, women—into electrical construction work and the union. The outward show of integration of black and Hispanic men and other minorities into the electrical brotherhood and the trade presents a paradox: given the prevailing racist climate in the country as a whole, why have African American and other minority men made inroads into the trade, and achieved relative acceptance in the industry and on work sites, while women have not? In addition, why did the electrical brotherhood, while downplaying and resisting women's entry, take political leadership on the integration of minority men? The role of the brotherhood's fraternal clubs and the ability of the civil rights movement to challenge racial bigotry help explain this paradox.

More broadly, social movements for equality are not generally factored into economic explanations of social change. To the extent that they examine them, economists look only at the degree to which legislation resulting from those movements can be linked to changes in labor market outcomes.[26] But social movements serve symbolic and informational goals in representing the grievances of a group of people. Thus, examining their role in achieving change is critical.

The antecedent movement for black equality in the 1950s and 1960s experimented with new campaigns to gain access to better jobs. Author Nancy MacLean (2006) documented these efforts on the part of the NAACP, which had labor and industry committees within its branches throughout the country.[27] As presented in Chapter 5, the year 1961 witnessed the growth of coalitions in New York City led by A. Philip Randolph which drew Malcolm X, then a Nation of Islam minister, into a joint rally with trade unionists and civil rights leaders demanding their fair share of jobs in the racially segregated building trades.[28] In a nutshell, driven by a desire for their rightful place at the table of skilled trades work, civil rights leaders such as Herbert Hill argued for voluntary agreements on the part of industry and craft union leaders for black men's equal opportunity in the construction workplace. He and other civil rights leaders in the early 1960s further argued that the white male bore the unique responsibility to restore the dignity and masculinity of the black man as male breadwinner and patriarch in the American nuclear family. (Anthropologists such as William Whyte, author of the renowned ethnography *Tally's Corner* and sociologist Daniel Patrick Moynihan have suggested that the devolution of black "manhood," robbed by the institution of slavery and lack of access to good-paying jobs, resulted in the disarray of the black American family and black matriarchy.) Civil rights leaders also sought to build alliances with building trades unions for the purpose of addressing poverty and inequality in the black community.[29] They argued, for example, that African American men have been a presence on construction sites since the late nineteenth century—sometimes as skilled craftsmen—but were never duly recognized. This classification stemmed from white tradesmen's desire to limit labor supply, as well as to a lingering tradition of racism in the skilled trades which labeled the most sophisticated and skilled black tradesmen as "common laborers" and thus making them ineligible for craft union membership.[30]

In the early part of the twentieth century, dangerous job sites and low wages for construction workmen meant less competition for jobs and a more racially diverse workforce.[31] As construction unions became more powerful and wages increased for skilled tradesmen, the jobs became more desirable and the industry became whiter; whites continue to occupy this privileged niche today.[32]

Nonetheless, the civil rights movement resulted in an overall increase in the number of black union members nationally. From 1979 to 2006, black workers made up 13 to 15 percent of all union members while the proportion of whites over the same period fell from 78.1 percent to 69.2 percent.[33] The rapid expansion of the building trades industry of the early 1960s (sometimes referred to in the industry as the "hurry up and build" period) and the concerted actions of black leaders influenced government enforcement of equal opportunity laws. The voluntary efforts of Local 3 and the fraternal character of the brotherhood initially helped black and other minority men to gain footing in the trade.

Given the laws of supply and demand, it is often assumed that people of color would have had a better chance of integrating into the electrician trade during periods of expansion. Paradoxically, it was during one of the building trades' most severe recessions that African Americans made their most significant inroads into the electrical brotherhood's apprenticeship program. In the stagnation period of 1974 to 1980, black and minority men made up 3.6 percent of the total union membership; during the economic boom of the mid- to late-1980s, their numbers actually decreased.[34] Despite this, the successful integration of blacks into the trade and industry paved the way for other racial minority men, as well as a handful of women.

Male racial minority groups—blacks, Asians, Latinos—have had different experiences in the union and trade. Their success or failure depends on their ability to develop an occupational enclave based on well-organized and effective informal referral networks in the industry. For example, the slow but steady progress of minority men into the industry since the 1980s, as discussed extensively in Chapter 5. Although the voluntary efforts on the part of the Local and especially Harry Van Arsdale Jr., its visionary leader, contributed to successful integration, they were a poor substitute for strong governmental enforcement, which precipitously weakened during the Reagan years of the 1980s.[35]

Nonetheless, the percentages of Hispanic workers in construction and electrical work in particular have increased significantly. Today, a typical construction site in New York City neighborhoods like Harlem includes immigrant and racial minority men mainly from Central and Latin America (e.g., Mexico, El Salvador, and Ecuador).[36] These workers are more likely than their white male counterparts to be nonunion, which further fuels racism on work sites.

Respondents report that graffiti on job sites expressing racial hatred has increased. But racism is not only black versus white. Xenophobia has also increased among workers as newly arrived immigrant groups are pitted against native whites and racial minorities.[37] Ethnic groups as well as white European American men believe they are losing ground to New York City's immigrant workers. For example, black workers, in particular, have made very small gains over the last decade. Yet women have made even fewer gains. Their participation remains marginal, despite the strong demand for skilled apprentice labor with the rebuilding of downtown Manhattan and an infusion of federal funds into the city for infrastructural work on roads, bridges, and buildings. Antidiscrimination legislation, weakly enforced if at all, has done little to remedy this situation.

In addition, the integration of black men into the electrical trade manifests similar problems that informal networks of referral and recruitment pose for women's inclusion in nontraditional jobs: they militate against women's hiring. Although father-to-son sponsorship is decreasing, male workers and unionists continue to guide male relatives and buddies into the trade. However, pressured by the threat of litigation brought by women pioneers in the 1980s, male unionists initiated some father-to-daughter sponsorships beginning in the 1990s whereby male tradesmen recruit their daughters into the electrical brotherhood.

On the political level, the complex linkage of political parties and legislators to contractors, industry leaders, and labor constituents hamstrings progressive unionists and lawmakers in their efforts to enforce sex desegregation orders. Present-day discrimination against women in the building trades proceeds unabated, either intentionally or through benign neglect.

Occupational Culture

When a handful of women electricians entered the construction division of Local 3, a tradesman described the reaction of men in the industry as "waving a red flag in front of a bull." There were six women in the first class of women electrical apprentices. What was all the fuss about? According to one male, "these women hit a nerve." The institutional "nerve" of craft trade union customs and traditions operates beneath the surface of the brotherhood's collective conscious to keep women "in their place" as electrician colleagues and "brothers" in the union.

Aside from these institutional roadblocks, male occupational culture on work sites can prove a formidable barrier to women becoming settlers in the trade instead of merely passersby. Male occupational culture is central to explanations of male workers' hostility toward women in nontraditional jobs in such public sector occupations as firefighting and policing and in private sector occupations like mining.[38] This case study of women electricians raises some additional questions regarding the extremely hostile culture women encounter at work sites: Why are men in the construction industry unwilling to tolerate changes in their work culture, or relinquish any control over their labor organizations? How does their resistance to women on the job site relate to gender relations at home and economic insecurity? In addition, what role does race play in men's reactions to women on the job site?

Male bonding and camaraderie existed prior to women's entry into the trades. Nonetheless, women's presence sharply reminds male workers that the value placed on physical aspects of electrical work are continuously undermined by new groups of less-valued workers, namely, women. Male workers often try to rid the work site of women workers by vulgar displays of pornography or inappropriate jokes or language. Tradesmen are fearful that sexual favors by women to male straw bosses will earn them bonus points for plum job assignments ("tit" jobs), promotions, and overtime. As decent work conditions in construction continue to deteriorate in the city, women workers are more likely than their male counterparts to demand safer and cleaner work sites. Male workers perceive these efforts to gain better working conditions as too demanding, and claim that contractors will turn to nonunion labor if pressured.

Gender discrimination and sexual harassment continue to be endemic to the construction trades and affect both black and white women. Black women and women of color generally try to cope as best they can in order not to become more conspicuous in a sea of male hostility. Regardless of such efforts, women often become scapegoats for the fears of white and minority male tradesmen. For black women, the violence and harassment on construction work sites is pandemic. Victoria King, an electrician journeywoman at the Jacob Javits Center in New York City, was beaten and raped. King quit the electrical trade in the 1990s as a result of this violent sexual assault and became a lawyer. Before launching her new legal career, she sued the contractor on the Javits job and has since written a book about her ordeal, *Manhandled Black Females*.[39]

Black tradeswomen generally are also fearful that men will retaliate if women raise issues regarding sexist attitudes. According to Donna Simms, an electrician journeywoman in New York City and a former member of an activist group:

> This is not a realm where I confront people on their conscious or unconscious hostility. In the construction industry, women of color do not have the critical mass to nail their associate to the wall for their casual racist attitudes, or sexism that they put out. It is also difficult to know where their racism ends and sexism begins. They both—racism and sexism—come from the same source, and I can't easily and clearly separate it. Just like torque, one compounds the other."[40]

If they are related to black men in the union, black women also fear that filing a formal complaint against another union member will lead to the blacklisting of their entire family. White women are more likely to formally complain about harassment, unless they also have family members in the trade.

Case studies of women coal miners, police officers, and truck drivers suggest that women respond differently to adverse work conditions than their male counterparts. But there are also some observable differences between white women and women of color in the building trades. Black women are more likely to put economic concerns before gender, though this does not imply they are any less harassed or discriminated against. This discrimination prompts black women to develop their own collectives, or to join in with black men in established fraternal clubs like the Latimer Society.

Their foremost concerns are dealing with hostility from tradesmen, improving conditions on the work sites, and learning the trade in order to obtain a journeyman's card.

The degree to which employers overlook gender discrimination makes it undeniable that this behavior is a conscious and systematic effort, mainly economic and cultural, by employers to discourage women from entering the industry.[41] In addition, the "go along to get along" mentality of male workers, their straw bosses, and union leaders is a silent but salient factor contributing to its ongoing practice by contractors. Just as in other private sector occupations, like mining, and in public sector occupations like policing and firefighting, male employers and unionists are often times unsympathetic to women's mistreatment on the job, and male workers can become vicious. "Intimidation doesn't even come close to how I felt, it was pure fear," said a woman mineworker.[42]

Across the building trades industry, women are an unwelcome incursion in an industry that still holds some opportunity for men (mostly white) with little or no higher education to move into the middle class. Economic downturns and the increasing growth of nonunion work are eroding this economic progress, and adds yet another dimension to men's resistance to women. Tradesmen resist women because they are generally fearful for their jobs, and their perceived notions of women's comparative sexual advantage in competing for plum assignments buttress this fear. However, large construction projects in New York are federally subsidized and contractors are required to comply with antidiscrimination laws in their bids for these lucrative building contracts. At the same time, these projects provide increased employment for tradesmen and benefit white male workers the most.

The following is a comment from a journeywoman on the pervasive sexual harassment in construction.

> It might begin as the act of a disturbed worker on the job. But when a gang of men does not stop the abuser, it appears to be sanctioned by the group. Sexual harassment becomes part of the ritual testing of female apprentices and reminds the women that they are extremely vulnerable and unprotected in an all-male environment. Female apprentices are made to feel that they have transgressed the boundaries of sexual morality and cannot, therefore, expect to be treated with common human decency.[43]

Some men even say that women use harassment charges to cover up their poor workplace performance. The owner of a small contracting business who serves on the Joint Apprenticeship Committee says, "Up here we often get sexual harassment job complaints when a female apprentice is brought before the termination committee. Usually, we will review her record and there will be high absenteeism. We have found that women will use the sexual harassment thing as a last-ditch effort to save themselves."[44] Even the most unpleasant work sites have to be staffed, so the work rules governing apprenticeship allow for rotation out of a difficult or uncomfortable job only once a year. For a woman being harassed, there may be no quick or simple way out, only the choice not to come to work at all.

Not all tradesmen are abusive toward women. Some are merely passive. Others still are unhappy about how women in construction are treated. But even these men, who could be allies to women, are generally reluctant to intervene when they witness harassment. They fear being ostracized by their coworkers and penalized by their supervisors if they defend or support women on the job site.

The union has a particular interest in keeping its construction division organized as something of a paramilitary organization. This model worked well in the past to control union members and to confront hostile employers and nonunion contractors. In addition, the "rotation of elites" among a handful of families through several generations of union leaders served to keep the union leadership basically all-male and mostly European American. For decade after decade, sexism was hierarchically reinforced. Blind obedience is the order of the day. One former journeyman turned contractor stated, "So here is what we achieved all these years and now I have some little girl telling me it's not good because it doesn't suit some type of Betty Crocker mentality." Further, Evan Ruderman observed: "The Local intimidates the foreman and this intimidation is passed down the line as a form of social control. Within the hierarchical chain, no one will speak up or show any type of personal courage, even in the face of things they know to be immoral." Perhaps these men feel trapped in a "code of silence" regarding women's treatment on the job.[45]

All the women interviewed for this book agree that the brothers who are the most hostile to them are middle-aged men who feel the squeeze of family pressures and expenses. These men are especially fearful of women electricians and do not want to incur union—or their foreman's—wrath by supporting them. Foremen have tremendous power since they make the work assignments and lay off the workers.

Men who do speak up for women on the job are usually over fifty-five and heading for retirement. At this point, they have nothing much to lose. The younger journeymen and male apprentices, however, often define their masculinity, especially on the job, in contrast to female characteristics. Like the men in the military who author Christine Williams (1992) described in her important work on men's reactions to women in the Marines, and armed with the same type of military mentality, electrical journeymen, foremen, and unionists all cultivate machismo.[46] Enduring hardship is part of the construction mystique—and helps the union achieve its aims of workers' discipline on job sites, obedience to union work rules, and loyalty to the union leadership.[47]

Reports on the growing number of fatalities and injuries in the construction industry paint a dangerous picture for construction workers today.[48] Veteran workers have commented that over the past twenty years, safety has declined on the job site. By tolerating worse conditions, nonunion companies have lowered the standards of job safety for the entire industry. Construction unions are often trapped in a situation where they feel they must discipline their workers not to complain about work site conditions for fear that a more laissez-faire nonunion contractor will be more competitive in his bidding. This mentality of enduring hardship among tradesmen adds to an ethos that encourages masculine resistance to women.

Men interviewed for this study, from employers to union leaders to rank-and-file workers, assert that women cannot expect to be treated equally if they demand different conditions on the job site and in their working lives. Similarly, in other occupations such as mining, when it comes to disciplining male union members for their harassment of women on the job, "that dirty work," a United Mineworker leader claimed, "is up to the company."[49]

The day-to-day harassment of women on construction sites serves as a warning to future entrants that they can expect the same treatment if they dare to enter a male-dominated preserve. Women are discouraged from applying for apprenticeships and are reluctant to enter this highly remunerated trade, even though today's new technology, such as the use of fiber optics on the construction site, means there is less need for physical strength on the job. This new technology has exacerbated gender tensions, with many male workers feeling their inherent advantage has been lost, leading them to actions ever more protective of the brotherhood.

Generational Differences, Dyke Baiting, and Workplace Acceptance

All too easily, women can become collaborators with their abusers in their own subordination. Women in blue-collar trades tolerate unjust treatment so as to gain acceptance from their fellow workers. Most female electricians seek the approbation of male workers, so the women become at once highly competent in their jobs while at the same time conforming to traditional female models. Some dress in a feminine manner to impress coworkers or superiors. Some flirt and tease. Even lesbian journeywomen may feel compelled to flirt and date men to hide their sexual orientation and avoid intimidation. Some women even grant sexual favors to male superiors. Journeywoman Melinda Hernandez has observed that women who entered the industry in the 1990s and today "use their sexuality to get over."

These tactics infuriate the pioneer generation of women construction workers. They, too, often felt powerless. But they remember their history of unity and militancy on issues important to them, and they remember that they were not reluctant to report instances of abuse to agencies outside the union. They know how stories about sexual availability can follow a woman from job site to job site, making sexual accommodation ultimately self-defeating. Pioneer Cynthia Long states that "the construction workplace is full of gossip, your reputation precedes you on job sites, whether you are a serious crafts-man, a 'stiff' who doesn't do any work, 'connected' meaning you have family in the industry, or in the case of women, you 'put out' for the men."[50]

"Respect among the men in the industry is paramount in craftsmen maintaining a good reputation," states Hernandez. "The thought of younger women coming in today and using their sexuality in exchange for favors or light duty enrages me."[51] The women who entered the trade via the feminist movement feel a certain animosity toward later entrants, who seem un-aware of the struggles of their older sisters. The "I'm not a feminist, but . . ." attitude of second- and third-generation women workers especially angers the pioneers. One of them recalls how hard they struggled: "Nothing was just given to you. If you didn't make it happen, it wasn't going to happen. If you didn't demand a bathroom, you wouldn't get one. If you didn't lay down the law, there was going to be no law." She explains the conflict with younger women this way: "We got a lot of flak for having such aggressive attitudes. We got called everything in the book. From this, we developed stigmas that we're still living with. And I think we developed chips on our shoulders as well."[52]

Cynthia Long observes:

As each class of women came in, we felt resentful because they were coming in and being more flippant. We felt they didn't really realize how difficult it was. A lot of women didn't feel it was necessary to band together, but what I think was really happening was their own fear. They were hearing their male coworkers talk about other women as being "real ball-breakers" or "real bitches." I think being really young and wanting to be accepted, some classes of women just tried to be nice. They were trying to be "good girls" and make themselves different from the "bitches" that came in the begin-ning. . . . [she laughs] You know, like me.[53]

Unlike the pioneers, the next waves to enter the trade—working-class white women and women of color—generally dealt with sexual harassment not collectively but individually. Those sponsored for membership by union relatives or boyfriends often did not acknowledge any problem at all. This is also true of the latest, youngest generation of women, who seem especially sensitive to the possibility that "dyke baiting" can be the consequence of any self-assertion on the job.

Even today, dyke baiting is endemic. One journeywoman claims: "You could be the most feminine woman on the job. But if they don't like you the men will still say you're gay. There is a constant discussion of which women are gay and which are straight."[54] Characteristics that can mark a woman as lesbian are assurance, assertiveness, physical strength, and any other notable difference. Surprisingly, many men are more threatened by working with gay women than by working with gay men. This may be because of the patriarchal ideology that women should occupy certain roles and not others. However (sexual politics can be complex!), some men can accept gay women in the workplace more easily than they can straight women. As one journeywoman explains, "If she's gay, the men can sort of understand why she's in a man's job." Most white men in the industry do not feel that they are discriminating. These contradictory attitudes are often derived from their own sense of place inside hierarchies of privilege at work, in the union, and based on experiences of their everyday lives in the family and society.[55]

Regardless, for many male workers, contractors in the construction trades, and union officials, women at the work site—gay or straight—are an unsettling presence. Clearly defined gender roles that also maintain male dominance in the family are an integral part of the construction trade culture, and this ideology prevents women from gaining the lucrative and stable jobs that are available in Division A. While working styles, technological advances, and the dangers of construction work contribute to the conflicts between male and female workers, the hostility of male contractors and unionists to women is the major reason for sex segregation in the construction building trades and the principal reason why Division A of Local 3 remains overwhelmingly male.

Fraternalism and Privilege at Work and at Home

Working-class tradesmen tend to be open and emphatic about their feeling that women do not belong in the industry. According to author W. J. Rorabaugh (*The Craft Apprentice*, 1988), tradesmen in earlier centuries were not only masters of their craft or over apprentices, but the "master" label automatically extended to their dominance over black men and the women in their lives. The "master" journeymen had implied patriarchal rights over his family. In eighteenth-century male fraternal societies, men were viewed as masters in the sphere of commerce, but unfit to guide the morality of the larger society.[56] Men came to symbolize the brutalization of the new market system and women, the symbol of public good and morality.

In the late nineteenth and early twentieth centuries, men in fraternal societies began to challenge this public role of women. For example, in societies like the Masons, men's social control over the norms of the broader society was restored while a women's public role was reduced to a private one inside the home.[57] Advanced capitalism brought about a harsh world of business practices and the further degradation of labor. In this environment, women inside the family came to symbolize a "haven in a heartless world." A tradesman's family and the woman in his life served to provide the emotional and material reproduction that permitted him to do battle in a brutalized labor market.[58] Despite changing gender relations in the larger society, and post-World War II alternative forms of family life and into the twenty-first century, the electrical brotherhood works incessantly in its organizational structure, work culture, recreational and commercial networks to maintain its gender-polarized definition of women's role and family life.

Male electricians' family status as primary wage earner, the brotherhood's reinforcement through social club rituals of this male breadwinner mentality, and the nuclear family all contribute to tradesmen's view of women in the trade as attacking their lifestyle and betraying the cause of the working man.[59] In addition, the invariable demands that women make for reproductive health care, and their identification more often than not with feminist issues such as child care and a discrimination-free work environment, make tradesmen especially antagonistic to women as colleagues.

Patriarchy influences both women and men. Clearly, men at the top are determined to maintain the hierarchical relationships that solidify their power, and brotherhood and solidarity help them to do this. Men at the lower echelons of the organization are themselves oppressed by patriarchal relations.[60] They respond in contradictory ways. The majority of apprentices and journeymen accept patriarchy and its concomitant oppression of women because of economic necessity, the fear of change, and the desire to maintain whatever privilege they possess within their own families. As we have seen, however, when male apprentices were faced with a crisis of their own, some began perceiving themselves as an underclass and enlisted female apprentices to act collectively with them. This suggests that in some instances, the traditional gender hierarchy can be overturned.

For most white male workers, women's integration into the trades and union is even more difficult to accept than that of minority men. The presence of women in the construction trade is perceived to be a threat to male workers' power and to their privilege in their homes. Perhaps women electricians threaten the manly image of construction workers forged at the beginning of the century and sustained in the face of continual job insecurity. Perhaps that is why men have made the work site so unpleasant for women.

At home, male electricians are more privileged than their female counterparts and this male gender privilege also complicates women's struggle to become "brothers." More often than not, especially in good economic times, a unionized male journeyman can financially support a "stay at home" wife and dependent children. Even young male apprentices anticipate not only inheriting the status of master electrician but also the privileged position it may bring as the breadwinner in the family. At every level, on the job site, at home and in the union, women electricians encountered a sexual hierarchy in the electrical brotherhood and trade.

The Sisters in the Brotherhood: Women Organizing Women

The District Council of Carpenters in New York City runs a program for tradeswomen called, the "Sisters in the Brotherhood." But like the electrical brotherhood within the union tensions loom large regarding tradeswomen's gender equality issues among the male membership, among the leadership, and among tradeswomen as well. The resistance of women into the ranks of brotherhoods in the building trades poses tactical dilemmas for women who are pro-labor feminists.

Feminism is viewed by many male unionists as antithetical to union discipline and union loyalty, the very practices that are at the core of the brotherhood's ascendancy to power. One journeyman in the electrical brotherhood put it like this: "Women shouldn't organize themselves or change the union as a feminist project. That doesn't belong here."[61]

Local 3's definitions of solidarity and brotherhood are based on a gender ideology that makes the integration of women into Division A more challenging than in other crafts, or even in other electrical construction locals. For example, an IBEW local of construction electricians in Seattle, Washington, has made greater progress in admitting women, and a smaller electricians' local in Long Island, Local 531, has a better track record in integrating women into its union.

While social clubs facilitate the integration of ethnic, racial, and national minorities into the Local, they present women with a formidable obstacle to their integration in the union and the trade.[62] Female members have a hard time making gains into the trade because of their marginalized position in this fraternity of the construction division. The causality is indirect: women are excluded from this fraternal bonding, thus isolated from the networks that confer power, thus kept subordinate.

Nonetheless, it is not uncommon for women in blue-collar nontraditional work to form women's clubs. Labor history in New York is replete with stories of working women's clubs at the turn of the twentieth century.[63] Both men and women have turned to a collective conscious mode of organizing in electrical work and in other occupations and industries as well. Judy Foster, a trucker, notes that her coworkers created an environment of support for one another so they could learn a male-dominated trade in a noncompetitive atmosphere.[64]

Men and women in the Local as well as at the level of the IBEW have created caucuses and internal clubs around aspects of identity politics in order to challenge the brotherhood's race and gender hierarchy, employers' hostility, and job insecurity. Ethnic minorities, such as Jews in the late 1920s, developed workers' caucuses to break into the trade. Men of color during the 1960s and 1970s developed worker collectives based on identity politics of race. In the present day, newly arrived immigrant workers from developing countries like India and parts of Latin America continue to organize on the basis of ethnicity, nationality, and race to gain a foothold in the industry and trade.

Women electricians and other tradeswomen from across a spectrum of the building trades have created a work culture based on gender solidarity, but women in the electrical trade do not represent a unitary category and their small numbers in the industry hamper their influence. Women electricians experience varying degrees of privilege at work, divided along lines of class, race, and their association with the feminist cause, as well as their links to informal networks such as male kinsmen.

Divisions run deep among women in the building trades on the whole.

Based on their class background, affiliation with the feminist cause, sexual orientation, and race identity, women electricians interviewed in this book would either join feminist collectives such as the Women Electricians club or union-sanctioned groups such as the Amber Light Society, but not both. In addition, sexual orientation and racial identity are strong factors that influence whether or not women will join lesbian rights groups across the trades, or traditional male ethnic and racial fraternities.

Even among "feminists" there are divisions. Liberal feminists among women electricians emphasize individual civil rights regarding harassment and pornography as the central focus of their collective conscious. These women entered the industry through feminist-type organizations such as All-Craft and Nontraditional Employment for Women. Progressive feminists view their position in the trade as a symbol of women's rights, and more closely identify with broader socialist movements for equality. They want to change the unequal gender hierarchy of male dominance in the brotherhood. Radical feminists place harassment and pornography as the central focus of women's efforts to organize, and often times alienate working-class women and women of color who view broader priorities such as inadequate training as a more pressing issue.

Although women electricians have attempted to form sororal clubs, first the Women Electricians club and later the Amber Light Society (ALS), they have found it almost impossible to get the electrical brotherhood to recruit more women into the apprenticeship and the industry. Thus women experience a revolving-door phenomenon; women enter and leave in small numbers and these small token numbers are then replenished. Since there is no emergent critical mass of women, the leadership and employers do not take their issues seriously on work sites. Women are generally left out of the power loop. Like their electrician sisters in the Women Electricians club, many members of ALS feel marginalized and excluded.

The situation of women of color is especially complex. Most trades-women who are progressive feminist types, as well as women of color recognize that black men opened the industry for them by fighting the early battles over segregation. The daughters, wives, and sisters of white tradesmen who are sponsored into the union, however, acknowledge a greater debt to the Local. Regardless of race, many younger women do not give enough credit to the women pioneers, minority or white, who struggled to make female entry into craft apprenticeship possible.

After the entry into the trade by the first group of pioneer women, subsequent generations of women electricians were increasingly connected by ties to family already in the Local. These women were fearful of associating with the "feminist types" in the women's club and developed alternative, fraternal organizations within the brotherhood that did not rock the boat.

On a positive note, both the pioneers and younger women are unified in their appreciation for the union's financial support for higher education. For some women, it is the reason they entered the industry. Women as well as men are supported in their efforts to earn associates and bachelor's degrees in labor studies. Some members of Local 3 have gone on to law or business school and have returned to serve the union. Others have pursued different professional degrees; one Hispanic woman eventually left the industry and became a pediatrician.

Getting the Bugs Out: Women's Place in the Trade and Brotherhood

A young male apprentice, newly recruited into the electrical program and interviewed at a focus group meeting, asked: "So, what will it take to change the attitude of contractors and the union to recruit more women into the trade?" Referring to one early twentieth-century jurisdiction of electricians, ridding dumbwaiters in New York brownstones of insects, a male veteran electrician in the same group responded: "It's just a matter of 'getting the bugs out.'" But women electricians are not finding it so simple. They are still struggling to organize a collective voice—one that includes their male colleagues—for social change within the industry and brotherhood.

The mere presence of women in construction work offers tangible evidence of the sexual revolution of the 1960s and 1970s. Female construction electricians have clearly freed themselves from a sex-defined subservient role. The majority of women in my sample were well aware that their presence in a nontraditional job overturned traditional expectations. They went forward nonetheless. This struggle to be accepted in a male-dominated occupation contradicts notions of women's passivity in the workplace. The tradeswomen pioneers and the subsequent waves of women who joined them worked to improve conditions for all construction workers.

Despite their frustration and political defeats, these tradeswomen represent an emerging culture of working-class feminists within the trade union movement that serves to challenge male privilege and power, redefine notions of a woman's work, and expand and transform our notions of brotherhood. Their efforts reinforce the meaning of solidarity and speak to the right of women to earn a living wage in all lines of work and to be treated equitably in workers' collective organizations.

Evan Ruderman remarked:

> It's not just about keeping on raising my wages and keeping on defending the right of skilled workers to have higher wages, but it is also related to the fact that I really do support a nurse's need to be paid more, a social worker's need to be paid more, and a teacher's need. A lot of service jobs that are predominantly female need better compensation so that we stop having this huge segregation in terms of pay and benefits. It has to become more of a situation where you choose a job because of what you really like rather than being forced to choose one because of economics.[65]

Many more women might choose jobs in the construction trades if they felt they were welcome. There is certainly no shortage of demand for skilled workers in New York City. And it is union jobs women want—not the exploitative and dangerous work that nonunionized contractors offer women, minorities, and immigrants. At the same time, women are discouraged from applying for union apprenticeships because of the famous hostility that awaits them on the job. This situation is true across the skilled trades—in carpentry, ironwork, and steam fitting. The contractors have hung "Men Only" signs on their doors and the unions have played along. Nepotism rules the hiring process in construction, and without union cooperation, women cannot get jobs.

The Bush administration was not much help. Title VII laws that prohibit gender-based discrimination and sexual harassment were simply not enforced. Laws that require contractors to make a "good faith effort" to recruit and retain women were ignored. The result is that affirmative action goals are not taken seriously either by contractors, who could force unions to send them female workers, or by the unions themselves, who do not even attempt to recruit more women. Ironically, it is these very constituents that must be brought into the brotherhood of labor in order to strengthen its position in this era, when the union's strength is steadily eroding. Indeed, with the help of nonunion contractors, the government is trying to wrest control of craft apprenticeships from the union, and the number of union jobs is dropping. The widespread use of nonunionized day laborers in general in the industry is indicative of the growing trend of union busting and the exploitation of more vulnerable workers, such as immigrants.[66]

Certainly government enforcement of Title VII would help. So would increased litigation. In the past, lawsuits have proven somewhat effective in influencing contractors and unionists to comply with the law.[67] Another possibility lies in training male unionists—foremen, shop stewards, and the leadership—to recruit and retain women in the trades. Presently, the Institute for Women and Work at Cornell University hosts a successful "journey-sister" program based on a similar program in the San Francisco building trades that helps seasoned journeywomen mentor new female apprentices. In addition, labor programs throughout the country as well as women's advocacy organizations like Federally Employed Women, Chicago Women in Trades, United Tradeswomen, New York City Tradeswomen, among many others, have provided sensitivity training and workshops for men on gender discrimination, sexual harassment, and illegal retaliatory action. Unfortunately, male contractors and unionists take little advantage of these learning opportunities. So far tradeswomen are interested in these programs, very few or no men. Perhaps the women are interested because they realize that they must take some steps—beyond relying on connected kin in the union—to infuse women workers' influence in the union movement, like the women electrician pioneers who paved the way for them.

Union-busting employers and contractors are exploiting already existing divisions of race, gender, and class to weaken the union sector in construction. Nonetheless, very few if any male unionists acknowledge that the union can gain strength by recruiting women to the brotherhood. Even some progressive policy makers have turned a deaf ear to the many hearings, public forums, and complaints of what is now three generations of tradeswomen.

On a note of optimism, the majority of women interviewed for this study find construction work to be liberating and exciting. They experience substantial job satisfaction in their craft and develop considerable self-confidence through the mastery of new skills and successful problem solving. One woman compares the experience of completing the apprenticeship to "the feeling of being an athlete." Work in the trades brings real independence to women and women electricians are understandably proud of wiring the intricate electrical innards of New York City. Their work, their responsibilities, the risks they take, and the salaries they earn are equal to those of their male colleagues.[68]

And there has never been a better time for New York employers and unions to address this thirty-year exclusion of women. Women have proven they are strong unionists who can unite with their brothers against nonunion contractors and can advocate effectively against managerial practices and governmental policies that lower the standards in the industry as a whole. There are women interested and more than ready to train for skilled trades work, especially young women of color and those in low-wage earning jobs with families to support. Furthermore, the New York job market for skilled unionized electricians and tradespeople in general is expanding due to efforts to "green" buildings, that is, retrofit already existing structures for environmental sustainability; and the creation of construction jobs due to the American Recovery and Reinvestment Act of 2009 (or what is more commonly termed the "stimulus package").[69] According to "Presidential Executive Order for the Use of Project Labor Agreements for Federal Construction Projects," signed into law on February 6, 2009 by President Obama, Executive Agencies are permitted to enter into prehire agreements with labor organizations (project labor agreements) on large-scale federally subsidized construction projects of $25 million or more.[70] Consequently, predictions for future job growth in the unionized construction sector remain positive especially in cities like New York with over 26,000 new construction jobs generated by the rebuilding of the World Trade Center site in lower Manhattan alone and $3.7 billion dollars expected in direct and indirect wages.[71] There are good-paying jobs and opportunities for economic independence and advancement, and satisfying work awaiting women workers in the skilled trades.

Epilogue: What of Our [Labor's] Future?

My hope is that this book will further inform public policy and workplace practices in the industry to help women escape the cycle of poverty associated with no-wage and low-wage earnings. Tradeswomen revolutionized our vision of "inclusion" and reordered values related to unquestioned assumptions of sex roles. They have shown great personal courage—risking safety and approval—to define themselves in opposition to "the hubris of the godmakers, the hubris of the male system-builders," that is, the influence of the system of patriarchy in the workplace.[72]

Nonetheless, companies and institutions—as well as individuals in general—rarely, if ever, concede power and privilege on their own. Accordingly, the road to economic recovery is already paved with "potholes" of unequal access. Ninety-three percent of the entire construction industry (including highway, bridge, and building construction) is male; 7 percent is female. The U.S. Department of Labor currently has an opportunity not only to realize the goal of "reconstruction" began by black civil rights leaders of the 1960s which aimed at racially desegregating the building trades, but also to move beyond the mere public relations hype surrounding gender equality in the industry as envisioned by the 1970s feminist movement. Accordingly, a Washington, DC–based oversight commission out of the U.S. Department of Labor should be established strictly for the enforcement of already existing federal and state regulations and statutes that require non-discrimination compliance by state and local agencies, contractors, and unions on federal construction projects. This commission, based on an evaluation of past discrimination, should develop new oversight procedures to ensure the preferential recruitment and hiring of women and minorities, as well as enforcement of minority- and women-owned small business contract "set asides" for federal construction projects.

Construction unions can help. There is no doubt that women and minorities need unions to negotiate with employers for better wages, decent working conditions, and protection of their human rights. Likewise, unions need to increase membership in order to successfully exercise their power and influence. At a meeting over twenty-five years ago in Los Angeles, an IBEW labor leader asked: "What of our [labor's] future? Each era writes its own history. Our union heritage, vibrant and strong, has been passed on to us. Where we go from here depends upon our Brothers and Sisters of today." Although many unions recognize the vibrancy and strength that women bring to the labor movement, their silence concerning the ill treatment of women and minorities in the construction industry can signify complicity, and speaks volumes regarding a labor organization's commitment to these constituents. Unfortunately, the open shop construction movement and even some female contractor associations, namely Women Construction Owners and Executives USA, have joined the anti-union chorus to vigorously oppose and challenge the use of project labor agreements (PLAs) by seizing on the exclusionary practices in the trades. Ironically, it is only within the unionized sector of construction which is subject to governmental regulations and compliance with affirmative action goals and quotas, that the greatest possibility lies for women and minorities to achieve equal opportunities to these good-paying jobs.

Finally, despite the complex challenges for women in particular to break into construction trades, there is reason to anticipate change in a positive direction. I argued in the Introduction of this book that culture is always temporal, contested, and emergent. Even the tight-knit male fraternities in the brotherhood cannot be explained simply as totally bought-off and co-opted. The various attempts of women electricians to form their own clubs such as WE, Amber Light, and others are also ways to open up a means to challenge a hierarchy of privilege and unfairness, and to voice the concerns of rank and file members. Furthermore, it is not the case that cooptation and conformity are the necessary outcome of activism in the brotherhood or for that matter, in any other male dominated institution. Cynthia Long, an Asian American veteran pioneer electrician, testifying at a human rights hearing in New York City, spoke to the possibility of change in the industry and trade. Long stated that with visionary and creative planning on the part of contractors and industry leaders, the doors for women could swing open in the building trades. Reflecting the optimism and perseverance of women in the trade, she cited the accomplishments in the 1980s of an electrical brotherhood in Seattle that inducted twenty-seven women (out of a class of one hundred) into its apprenticeship program. Although this rate of inducting women did not last, it illustrates that under the right circumstances, tradesmen can expand paternalistic notions of "unselfish" brotherhood to include craft sisters on a level playing field.

Nonetheless, it will take the type of personal courage women electricians and other tradeswomen have shown, moral courage on the part of their male coworkers and supervisors to speak out on their behalf, and much pressure—legislative, judicial, political, and educational—to reach farther than our current grasp for numerically meaningful inclusion, to push those doors open so that women can get the unionized construction jobs they want and deserve.

Appendix A

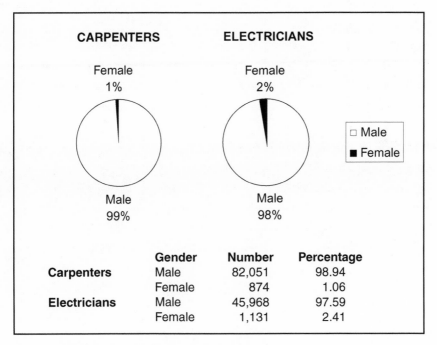

Fig A.1   American Community Survey 2006 Data

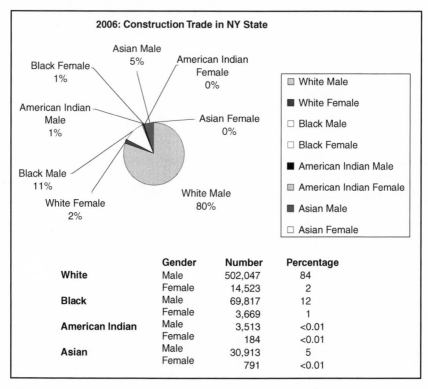

Fig A.2   B24010. Sex by Occupation for the Civilian Employed Population 16 Years and Over (*Source: Data set and survey: 2006 American Community Survey*)

Appendix B

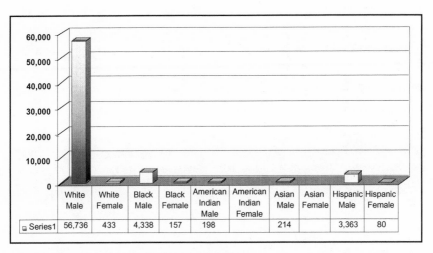

	White Male	White Female	Black Male	Black Female	American Indian Male	American Indian Female	Asian Male	Asian Female	Hispanic Male	Hispanic Female
Series1	56,736	433	4,338	157	198		214		3,363	80

Fig B.1 Carpenters in New York State, 1980 *(Table 217. "Detailed Occupation of the Experienced Civilian Labor Force and Employed Persons by Sex." (New York: New York State Census): 34-401–34-411.)*

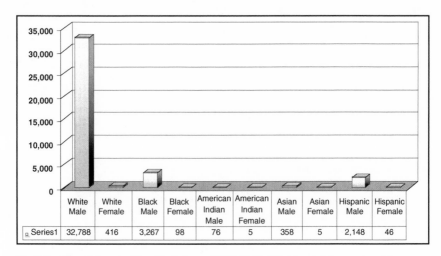

Fig B.2   Electricians in New York State, 1980 *(Table 217. "Detailed Occupation of the Experienced Civilian Labor Force and Employed Persons by Sex." (New York: New York State Census): 34-401–34-411.)*

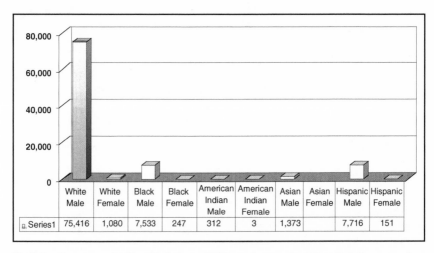

Fig B.3   Carpenters in New York State, 1990 *("Labor Force by Federal EEO Occupational Group" (New York: New York State Department of Labor). Available at www.labor.state.ny.us/ workforceindustrydata/eeo_pr2.asp?reg=nys&geog=01000036New%20York%20State.)*

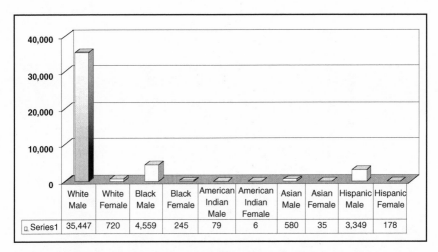

Fig B.4 **Electricians in New York State, 1990** (*"Labor Force by Federal EEO Occupational Group" (New York: New York State Department of Labor). Available at www.labor.state.ny.us/ workforceindustrydata/eeo_pr2.asp?reg=nys&geog=01000036New%20York%20State.*)

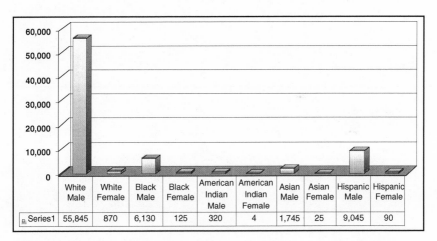

Fig B.5 **Carpenters in New York State, 2000** (*"Labor Force by Federal EEO Occupational Group" (New York: New York State Department of Labor). Available at www.labor.state.ny.us/ workforceindustrydata/eeo_pr2.asp?reg=nys&geog=01000036New%20York%20State.*)

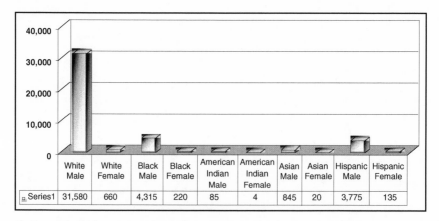

	White Male	White Female	Black Male	Black Female	American Indian Male	American Indian Female	Asian Male	Asian Female	Hispanic Male	Hispanic Female
Series1	31,580	660	4,315	220	85	4	845	20	3,775	135

Fig B.6   Electricians in New York State, 2000 (*"Labor Force by Federal EEO Occupational Group" (New York: New York State Department of Labor). Available at www.labor.state.ny.us/ workforceindustrydata/eeo_pr2.asp?reg=nys&geog=01000036New%20York%20State.*)

Appendix C

TABLE C.1   LABOR PARTICIPATION RATES IN CONSTRUCTION TRADES ACCORDING TO GENDER, TRI-STATE CENSUS DATA

DECADES	MEN	%	WOMEN	%	TOTAL
1970	206,500	98.8%	2,600	1.2%	209,100
1980	210,000	98.5%	3,300	1.5%	213,300
1990	286,132	98.1%	5,488	1.9%	291,620
2000	485,980	98.0%	9,862	2.0%	495,842

TABLE C.2   LABOR PARTICIPATION RATES IN CONSTRUCTION TRADES ACCORDING TO ETHNICITY, TRI-STATE CENSUS DATA

CENSUS YEAR	WHITE	BLACK	HISPANIC	ASIAN	OTHER	TOTAL
1970	198,400 (94.9%)	10,300 (4.9%)	n.a.	100 (0%)	300 (0.1%)	209,100
1980	191,400 (85%)	16,200 (7.2%)	11,900 (5.3%)	600 (0.3%)	5,100 (2.3%)	225,200
1990	245,014 (74.5%)	29,344 (8.9%)	37,240 (11.3%)	4,774 (1.5%)	12,488 (3.8%)	328,860
2000	371,011 (67.5%)	66,286 (12.1%)	99,491 (18.1%)	13,227 (2.4%)	n.a.	550,015

TABLE C.3A   LABOR PARTICIPATION RATES IN CONSTRUCTION TRADES
ACCORDING TO GENDER AND ETHNICITY, TRI-STATE CENSUS DATA

CENSUS YEAR	White		Black		Hispanic	
	MALE	FEMALE	MALE	FEMALE	MALE	FEMALE
1970	196,000	2,400	10,100	200	n.a	n.a
	(98.8%)	(1.2%)	(98.1%)	(1.9%)		
1980	188,600	2,800	15,800	400	11,400	500
	(98.5%)	(1.5%)	(97.5%)	(2.5%)	(95.8%)	(4.2%)
1990	240,856	4,158	28,350	994	36,498	742
	(98.3%)	(1.7%)	(96.6%)	(3.4%)	(98%)	(2%)
2000	364,560	6,451	63,563	2,723	97,876	1,615
	(98.3%)	(1.7%)	(95.9%)	(4.1%)	(98.4%)	(1.6%)

TABLE C.3B   LABOR PARTICIPATION RATES IN CONSTRUCTION TRADES
ACCORDING TO GENDER AND ETHNICITY (AS A % OF TOTAL), TRI-STATE
CENSUS DATA

CENSUS YEAR	WHITE MEN	WHITE WOMEN	BLACK MEN	BLACK WOMEN	HISPANIC MEN	HISPANIC WOMEN
1970	196,000	2,400	10,100	200	n.a.	n.a
	(93.9%)	(1.2%)	(4.8%)	(0.1%)		
1980	188,600	2,800	15,800	400	11,400	500
	(85.9%)	(1.3%)	(7.2%)	(0.2%)	(5.2%)	(0.2%)
1990	240,856	4,158	28,350	994	36,498	742
	(77.3%)	(1.3%)	(9.1%)	(0.3%)	(11.7%)	(0.2%)
2000	364,560	6,451	63,563	2,723	97,876	1,615
	(67.9%)	(1.2%)	(11.8%)	(0.5%)	(18.2%)	(0.3%)

TABLE C.4A   LABOR PARTICIPATION RATES ACCORDING TO INDUSTRY, GENDER, AND ETHNICITY, NEW YORK STATE CENSUS DATA*

Electricians

				Non-Hispanic Minorities				
CENSUS YEAR	ETHNICITY GENDER	TOTAL	TOTAL MINORITY	BLACK	AMERICAN INDIAN, ALASKAN NATIVE	ASIAN / NHOPI	MORE THAN ONE RACE	HISPANIC
1980	Male	38,637 (97.5%)	5,849 (14.9%)	3,267 (8.3%)	76 (0.2%)	358 (0.9%)	n.a	2,148 (5.5%)
	Female	570 (2.5%)	154 (0.4%)	98 (0.2%)	5 (0%)	5 (0%)	n.a	46 (0.1%)
1990**	Male	44,014 (97.4%)	8,567 (18.9%)	4,559 (10.1%)	79 (0.2%)	580 (1.3%)	n.a	3,349 (7.4%)
	Female	1,184 (2.6%)	464 (1.2%)	245 (0.5%)	6 (0.2%)	35 (0.1%)	n.a	178 (0.4%)
2000	Male	41,283 (97.5%)	9,703 (22.9%)	4,315 (10.2%)	85 (0.2%)	845 (2%)	679 (1.6%)	3,775 (8.9%)
	Female	1,069 (2.5%)	409 (1%)	220 (0.5%)	4 (0%)	20 (0%)	30 (0.1%)	135 (0.3%)

*"Labor Force by Federal EEO Occupational Group." New York State Department of Labor (June 2007)
**Electricians (except apprentices)

TABLE C.4B   LABOR PARTICIPATION RATES ACCORDING TO
INDUSTRY, GENDER, AND ETHNICITY, NEW YORK STATE CENSUS DATA*

Carpenters

CENSUS YEAR	ETHNICITY GENDER	TOTAL	TOTAL MINORITY	Non-Hispanic Minorities				HISPANIC
				BLACK	AMERICAN INDIAN, ALASKAN NATIVE	ASIAN / NHOPI	MORE THAN ONE RACE	
1980	Male	64,849 (99.0%)	8,113 (12.4%)	4,338 (6.6%)	198 (0.3%)	214 (0.3%)	n.a	3363 (5.1%)
	Female	670 (1.0%)	237 (0.35)	157 (0.2%)	n.a	n.a	n.a	80 (0.1%)
1990**	Male	92,350 (98.4%)	16,934 (18.03%)	7,533 (8.0%)	312 (0.3%)	1,373 (1.5%)	n.a	7,716 (8.2%)
	Female	1,481 (1.6%)	401 (0.4%)	247 (0.3%)	3 (0.0%)	n.a	n.a	151 (0.2%)
2000	Male	74,905 (98.5%)	19,060 (25.1%)	6,130 (8.1%)	320 (0.4%)	1,745 (2.3%)	1,810 (2.4%)	9,045 (11.9%)
	Female	1,128 (1.5%)	258 (0.3%)	125 (0.2%)	4 (0%)	25 (0%)	14 (0%)	90 (0.1%)

*"Labor Force by Federal EEO Occupational Group." New York State Department of Labor (June 2007

**Carpenters (except apprentices)

Appendix D

OVERVIEW OF THE CONSTRUCTION INDUSTRY

The construction industry focuses its work on building new structures and on additions, alterations, and repairs. The Department of Labor divides the industry into three major segments: general building contractors, heavy construction contractors, and special trade contractors. General building contractors build residential, industrial, commercial, and other buildings. Heavy construction contractors build sewers, roads, highways, bridges, tunnels, and other large projects. Special trade contractors are engaged in activities such as carpentry, painting, plumbing, and electrical work.

Construction work is usually coordinated by general contractors, who specialize in either residential or commercial buildings. They take full responsibility for the completion of the job. Although some general contractors have their own crews, they often subcontract the work out to heavy construction or special trade contractors. The special trade contractors typically do the work of only one trade, such as those mentioned above. Other than their own trade, they have no responsibility for the structure as a whole.

Working Conditions

Most construction workers work full time, with many working overtime. In 2008, more than one in five wage and salary construction workers worked 45 hours or more per week (Bureau of Labor Statistics [BLS], July 2008).

Construction workers must be physically fit. Work often requires prolonged standing, bending, stooping, and heavy lifting. Exposure to extreme weather is common because much of the work is done outside or in partially completed structures. Workers may be required to use potentially dangerous tools and equipment, often on temporary scaffolding or in bad weather. Therefore, construction workers are more prone to injury than people in other occupations.

Employment in the Construction Industry

Construction is one of the largest industries in the United States. Construction also maintains the most consistent job growth. About 64 percent of wage and salary jobs in construction were in the specialty trades: primarily plumbing, heating, air conditioning, electrical trade, and masonry. Twenty-four percent of the jobs were in residential and nonresidential construction; the rest were in heavy and civil engineering construction. In 2008, there were 7.7 million wage and salary jobs and 1.9 million self-employed nongovernment jobs (BLS, July 2008). Almost two-thirds were with general building contractors. The rest were with road and other heavy construction contractors. Construction jobs are generally concentrated in industrialized and heavily populated areas.

Employment in this industry is distributed geographically in much the same way as the U.S. population. There were about 883,000 construction establishments in the United States as of July 2008: 268,000 were general contractors; 64,000 were heavy and civil engineering construction or highway contractors; and 550,000 were specialty trades contractors, with the vast majority (65 percent) employing fewer than five workers. About 11 percent of construction workers are employed by these small contractors (BLS, July 2008).

Most construction workers are skilled crafts workers or laborers, helpers, and apprentices. The BLS classifies crafts workers into three categories: structural, finishing, or mechanical. Structural workers include carpenters, operating engineers, bricklayers, cement masons, stonemasons, and reinforcing metal workers. Finishers include lathers, plasterers, marble setters, terrazzo workers, carpenters, ceiling installers, drywall workers, painters, glaziers, roofers, floor coverers, installers, and insulation workers. Mechanical workers include plumbers, pipefitters, electricians, sheet metal workers; and heating, air-conditioning, and refrigeration technicians.

Training

Many workers enter the industry via apprenticeship programs. These programs offer hands-on training under supervision. Others enter the industry from less-skilled jobs such as helper or laborer. They acquire skills as they work. Other workers in the construction industry, not considered craftspeople, operate machinery and other construction equipment, including graders, bulldozers, scrapers, and paving, surfacing, and tamping equipment.

People may enter the industry without any formal classroom training after high school. However, skilled workers such as carpenters either need many years of on-the-job training, or apprenticeship training where they pick up their skills by working alongside experienced workers. As they acquire these skills they progress to more challenging work. After several years, they reach journey level. Apprenticeship programs are often administered by local unions. Training usually lasts between three and five years and consists of both classroom and real work experience.

The educational background in the construction industry is varied. High school graduates usually start out as laborers, helpers, or apprentices. Those who enter the industry from technical school may also go through apprenticeship training, but may complete it at a faster rate. Executive, administrative, and managerial employees usually have a college degree or many years of experience. College graduates often start as management trainees or assistant managers.

Earnings

Earnings in construction are higher than the average for all industries. In July 2007, production or nonsupervisory workers in construction averaged $19.53 an hour, or about $40,620 per year (BLS, July 2008). In New York, average earnings are $27.22 an hour (BLS, July 2008). Those working for special trade contractors earned somewhat more than those working for general or heavy construction contractors. Earnings vary according to the education and experience of the worker and the type of work, the size and nature of the project, the geographic location, and economic conditions. Earnings are also affected by the weather. Winter is a slack period for the

TABLE D.1. DISTRIBUTION OF WAGE AND SALARY EMPLOYMENT IN CONSTRUCTION BY INDUSTRY, 2006 (EMPLOYMENT IN THOUSANDS)

INDUSTRY	EMPLOYMENT	PERCENT
Construction, total	7,689	100.0
Construction of buildings	1,806	23.5
Residential building	1,018	13.2
Nonresidential building construction	789	10.3
Heavy and civil engineering construction	983	12.8
Utility system construction	426	5.5
Highway, street, and bridge construction	349	4.5
Land subdivision	97	1.3
Other heavy and civil engineering construction	112	1.5
Specialty trade contractors	4,900	63.7
Building equipment contractors	2,006	26.1
Foundation, structure, and building exterior contractors	1,132	14.7
Building finishing contractors	1,036	13.5
Other specialty trade contractors	726	9.4

Source: Bureau of Labor Statistics, www.bls.gov/oco/cg/CGS003.htm#earnings

TABLE D.2. PERCENTAGE OF WAGE AND SALARY WORKERS IN CONSTRUCTION CRAFT OCCUPATIONS EMPLOYED IN THE CONSTRUCTION INDUSTRY, 2006

OCCUPATION	PERCENT
Cement masons, concrete finishers, and terrazzo workers	91.7
Insulation workers	91.4
Structural iron and steel workers	84.8
Plasterers and stucco masons	81.6
Roofers	77.1
Drywall installers, ceiling tile installers, and tapers	75.7
Brickmasons, blockmasons, and stonemasons	71.8
Pipelayers, plumbers, pipefitters, and steamfitters	71.4
Glaziers	67.7
Electricians	67.6
Carpenters	56.8
Painters and paperhangers	46.7
Carpet, floor, and tile installers and finishers	44.1

Source: Bureau of Labor Statistics, www.bls.gov/oco/cg/CGS003.htm#earnings

TABLE D.3. MEDIAN HOURLY EARNINGS OF THE LARGEST OCCUPATIONS IN CONSTRUCTION, MAY 2006

OCCUPATION	CONSTRUCTION OF BUILDINGS	HEAVY AND CIVIL ENGINEERING CONSTRUCTION	SPECIALTY TRADE CONTRACTORS	ALL INDUSTRIES
Construction managers	$34.59	$36.90	$35.54	$35.43
First-line supervisors/managers of construction trades and extraction workers	26.23	25.96	25.77	25.89
Plumbers, pipefitters, and steamfitters	20.82	19.15	20.45	20.56
Electricians	19.62	20.17	20.45	20.97
Operating engineers and other construction equipment operators	18.29	18.90	18.29	17.74
Carpenters	18.07	17.97	17.50	17.57
Cement masons and concrete finishers	16.29	15.94	15.75	15.70
Painters, construction and maintenance	15.19	14.67	14.67	15.00
Construction laborers	13.15	13.24	12.60	12.66
Office clerks, general	11.03	11.08	11.02	11.40

Benefits and union membership: About 15 percent of construction trades workers were union members or covered by union contracts, compared with 13 percent of workers throughout private industry. In general, union workers are paid more than nonunion workers and have better benefits. Many different unions represent the various construction trades and form joint apprenticeship committees with local employers to supervise apprenticeship programs.

Source: Bureau of Labor Statistics, www.bls.gov/oco/cg/CGS003.htm#earnings

industry in colder parts of the country. Heavy rain may also delay a project. Earnings in selected occupations in construction in 2006 are in Table 1. About 15 percent of construction trades workers are union members or were covered by union contracts, compared with 13 percent throughout private industry.

Outlook for the Construction Industry

According to the BLS, employment of wage and salary jobs in the construction industry is expected to grow about 10 percent through the year 2008. This growth is actually slower than average for all industries, which is 11 percent. Between July 2008 through 2016, employment growth is projected to add approximately 550,000 new jobs in construction.

Employment depends mostly on the level of construction and remodeling activity. New construction generally decreases during periods of economic stagnancy or recession, and the number of available jobs in the industry fluctuates according to demand.

Employment in residential construction is expected to grow slowly because the projected slowing of population growth and household formation should reduce the demands for new housing. Additionally, the aging of the population will decrease the demand for larger single-family homes. Employment in nonresidential construction will also suffer from slow growth because the demand for commercial buildings will be decreased due to technological trends such as telecommuting, Internet shopping, home offices, and teleconferencing, as well as downsizing, temporary employees, and inventory reduction. Industrial construction is expected to be stronger because of the increased exports of the manufacturing industry. Old industrial plants and a large number of structures will have to be replaced or remodeled. Nursing homes and other extended care facilities will need to be built. Construction of schools is expected to increase to accommodate the children of the baby-boom generation.

Employment in heavy construction is expected to increase about as fast as the industry average. Bridge construction is expected to increase the fastest due to the need for bridge repair. Poor highway conditions will also increase the need for highway maintenance.

BLS projects that employment in special trades, the largest section of the industry, should grow slightly faster than the industry as a whole. Demand for special trades is rising. Remodeling should be the fastest-growing sector of the housing industry because of the number of aging homes.

Although employment in the construction trades is expected to grow as fast as the industry average, the rate of growth will vary among different trades. Employment of bricklayers, electricians, sheet metal workers, painters, and heating, air conditioning, and refrigeration technicians should grow faster than the industry average. Employment of carpenters, cement masons, concrete finishers, and terrazzo workers, plumbers, and structural metal workers is expected to grow more slowly than average because the demand for these workers will be offset by a greater use of new materials and equipment.

Many industry sources feel job opportunities are excellent in the construction trades because there is a shortage of skilled trades workers in general. The creation of construction jobs resulting from the 2009 economic stimulus package and the efforts of the Obama administration to promote green construction and revamp energy sources will heighten this shortage through 2016 because the pool of young people available to enter training programs is increasing slowly, and many in that group are reluctant to seek employment in the construction trades.

OVERVIEW OF ELECTRICIANS

Electricians bring electricity into homes, factories, and businesses, and to streets and roads. They install, maintain, and repair electrical wiring, equipment, and fixtures. They ensure that building construction and maintenance work is in accordance with relevant codes. Electricians may also install or service street lights, intercom systems, or electrical control systems.

Most electrical work starts with electricians reading blueprints. These blueprints show where circuits, outlets, panel boards, and other equipment are located throughout a building. When accomplishing their wiring tasks, electricians must adhere to strict national, state, and local building codes. Electricians join wires in boxes with various specially designed connections. Their hand tools are conduit benders, screwdrivers, pliers, knives, hacksaws, and wire strippers, as well as power tools such as drills and saws. Electricians also use altimeters, ohmmeters, voltmeters, oscilloscopes, and other equipment to test connections and ensure the compatibility and safety of components. (For a more detailed description of work performed by electricians, go to the BLS web site at www.bls.gov/oco/reprints/ocor015.pdf.)

Electricians also advise management regarding the operation of equipment and its potential dangers. They usually consult with engineers, engineering technicians, line installers and repairers, or industrial machinery mechanics and maintenance workers. Work is performed either indoors or outdoors. Electricians may have to do some strenuous work, as outlined above, but technological advances have decreased the amount of heavy work. They work a standard 40-hour week and often work overtime.

Electricians learn their trade through a union-sponsored apprenticeship program, which requires 2,000 hours of on-the-job training and 144 hours of classroom instruction. Electrical apprentices are under the supervision of joint committees in local unions in the International Brotherhood of Electrical Workers, and local chapters of the National Electrical Contractors Association. Programs are offered through union apprenticeship programs and private and vocational technical training high schools and community colleges. Most states require licensing of electricians; requirements vary from state to state. In order to become a master mechanic (journeyman), electricians must pass an examination that tests their knowledge of electrical theory and of national, state, and local building codes. Electrical contractors as compared with journeymen must obtain a special license based on more extensive knowledge of the trade.

There are approximately 624,560 electricians nationally in the industry and projected employment outlook should reach over 700,000 by 2016. The industries with the highest levels of employment for electricians are building equipment contractors, local government, employment services, nonresidential building construction, and electric power generation, transmission, and distribution.

Earnings

As of July 2008, the national mean hourly wage of electricians is $23.12 and their mean annual salary is $48,100 (BLS, July 2008). The mean hourly wage in New York is $30.09 (BLS, July 2008). Apprentices make on average 40 to 50 percent of what journeymen earn. Apprenticeship is typically four to five years of training but in some instances, such as electrical apprenticeship in New York, it can be extended to as long as six years.

The top-paying industries for electricians are natural gas distributors, nonresidential building construction, and the performing arts. The highest concentration of employment for electricians is in the South and West, and in Alaska. But the top earning states are New York, New Jersey, and California, followed by Alaska, Illinois, and Hawaii. In particular, San Francisco and the New York City metropolitan area (both heavily unionized) boast the highest earnings for electricians. In the Northeast, the mean annual salary of electricians is approximately $63,000 (*Occupational Employment Statistics, Electricians*, BLS, May 2007).

New York State has 34,000 electricians, and New Jersey is home to 16,000. In New York, electricians have a mean hourly wage of $33.79, and a mean annual wage of $72,000. They are among the highest-paid workers in the nonresidential construction trades, ranking second only to operating engineers and structural iron and steel workers.

On the lowest end of the earnings spectrum are construction laborers, whose mean hourly wage in New York is $23.96. Construction laborers are mostly newly arrived immigrants, and men and women of color. Other trades, such as carpentry, earn far less than electricians. For example, the mean hourly wage of carpenters is $27.76 (*National, State, Metropolitan and Nonmetropolitan Occupational Employment and Wage Estimates*, BLS, May 2007).

Notes

ACKNOWLEDGMENTS

1. *Lo Re v. Chase Manhattan Bank, N.A.*, described in a journal article by Nancy MacLean, "The Hidden History of Affirmative Action: Working Women's Struggles in the 1970s and the Gender of Class," The Journal of Feminist Studies, Vol. 25, 1999 (the Vladeck firm represented women officers, managers, and professionals employed by the defendant bank in a class wide discrimination action. Settlement was approved by the federal district court); also see, *Kyriazi v. W. Elec. Co.* 544 F.2d 1196, 13 Fair Empl. Prac. Cas. 1352, 12 Empl. Prac. Dec. P 11, 232; *Western Electric Company, Incorporated, Petitioner v. honorable Herbert J. Stern, United States District Judge for the District of New Jersey, Nominal Respondent, Kyriazi Cleo Kyriazi Plaintiff-Respondent*, No. 76-2044, U. S. Court of Appeals, 3d Cir. (argued Sept. 13, 1976, decided Nov. 5, 1976). The Kyriazi cases involved claims of systemic discrimination against all women employed by defendants. The judgment against the defendant and the settlement of all claims led to equal employment opportunity practices throughout the Bell system.

2. Evan Ruderman's theatrical monologue regarding her experience with AIDS was a major part of the documentary "Sex and Other Matters of Life and Death" shown on public television in 1996.

3. William Roseberry, *Anthropologies and Histories: Essays in Culture, History, and Political Economy* (New Brunswick, NJ: Rutgers University Press, 1994).

4. E. J. Hobsbawm, *Primitive Rebels: Studies in Archaic Forms of Social Movements in the 19th and 20th Centuries* (New York: Norton Library, 1965).

INTRODUCTION

1. Nancy MacLean, *Freedom Is Not Enough: The Opening of the American Workplace* (Cambridge: Harvard University Press), 2006.

2. U.S. Bureau of the Census, *Census of Population and Housing, 2000* (Washington, DC: Author).

3. MacLean, *Freedom Is Not Enough*.

4. The Javits Convention Center in New York City is a "plum" assignment in the building trades, and workers compete for jobs there. Because building projects at the center are federally subsidized, contractors have to show a "good faith" effort to hire women and minorities. However, racial conflict and alleged discrimination has resulted in a class action suit brought by minority and women plaintiffs against Javitts Center builders and owners. For the most recent case, see *Cokely et al. v. New York Convention Center Operating Corporation et al.*, Civil Action No. 00 Civ. 4637 (DAB) (AJP).

5. The Boston area has been a hub of tradeswomen's organizing since 1980. One of the first tradeswomen's organizations, Tradeswomen, Inc., was established in Boston to provide networking, information, and coping strategies for women in the construction trades. For detailed information on the organization, see Jane LaTour's *Sisters In The Brotherhoods: Working Women Organizing For Equality in New York City* (New York: Palgrove Macmillan, 2008).

6. I would not see this massive hole, the foundation of the WTC, again until after the September 11th attacks.

7. Peter J. Brennan was a Depression-era house painter who rose through the ranks of the construction trades to become the head of the Building Trades' Council, an amalgam of building trades' unions across a broad spectrum of craft trades. Brennan was a prominent labor leader for over three decades in New York and nationally. He was President Nixon's Secretary of Labor from 1973 to 1975, and is widely credited as the chief architect of the "hard hat" coordinated violence against antiwar peace demonstrators in lower Manhattan. He died in 1996; his obituary can be found online at: http://query.nytimes.com/gst/fullpage.html?res=9F05E0DD123FF937A35753C1A960958260.

8. U.S. Bureau of the Census, *Census of Population and Housing, 2000* (Washington, DC: Author); American Community Survey Data Set, *Civilian Employed Population 16 Years and Over,* U.S Census Bureau, Washington, DC), 2006).

9. See Appendix D for a full discussion of the types of building trades' apprentices and labor projections for employment; also see, U.S. Department of Labor, "List of Federally Funded Grants under the WANTO Grant Award to Women's Advocacy Organizations Providing Pre-apprenticeship Training into One Skilled Construction Trade Nationally," accessed January 15, 2007, at www.dol.gov/wb/03awards.htm. For a detailed discussion of the creation of green building strategies in the construction industry, see Robin Spence and Helen Mulligan, "Sustainable development and the construction industry," *Habitat International* 19, no. 3 (1995): 279–292.

10. Joseph M. Kelly, "Challenges and Opportunities: 2008 Profile of the Electrical Contractor," *Electrical Contractor Magazine*, July 2008; also for an informative discussion regarding the effects of the American Recovery and Reinvestment Act of 2009 on electrical building work, see also the feature article, "Recharging the Industry," *Electrical Contractor Magazine*, April 2008.

11. Irene Padavic and Barbara Reskin, *Women and Men At Work* (Thousand Oaks, CA: Pine Forge Press, , 2002), 20.

12. Betty Friedan, *The Feminine Mystique* (New York: Penguin Books, 1992).

13. Padavic and Reskin, *Women and Men at Work*, 22. See also Arlene Skolnick, *Embattled Paradise: The American Family in an Age of Uncertainty* (New York: Basic Books, 1991); MacLean, *Freedom Is Not Enough*; Lori L. Reid, "Devaluing Women and Minorities: The Effects of Race/Ethnic and Sex Composition of Occupations on Wage Levels," *Work and Occupations* 25 (1998): 511–536.

14. There have been many crane collapses and explosions at construction work sites in New York City which have resulted in worker fatalities, see, for example,

www.cnn.com/2008/US/05/30/crane.collapse/index.html; "Top Ten Most Dangerous Jobs" at http://ny-law-firm.com/top-10-most-dangerous-jobs.

15. On July 31, 2007 the National Labor Relations Board (NLRB) issued a decision that significantly impacts how union contracts may be negotiated for construction work. Prior to this NLRB decision, nonunion contractors and subcontractors were excluded from unionized constructions sites. The new decision prohibits owners, construction managers, and general contractors from agreeing with unions to require contractors or subcontractors to abide by union contracts.

16. Nonunion contractors often misclassify workers in the construction industry as independent contractors in order to avoid compliance with federal and state labor standards which require contractors—unionized and nonunion—to pay prevailing wages to workers on federally subsidized building projects; see also regulations of the Davis-Bacon Act at the U.S. Department of Labor Department, Labor Standards Administration, Wage and Hour Division, at www.gpo.gov/davisbacon/referencemat.html.

17. For example, unionized electricians in Local 3 initially resisted contractors' use of pre-fabricated switch and control boards on New York City work sites but eventually accepted the practice; see Lester Velie, "The Union That Gives More to the Boss," *Reader's Digest* (1956): 1–5; for a historical overview of attempts to deskill craft work, see David Montgomery, *The Fall of the House of Labor: The Workplace, the State, and American Labor Activism, 1865–1925* (Cambridge: Cambridge University Press, 1984); Michael Kazin, *Barons of Labor: The San Francisco Building Trades and Union Power in the Progressive Era* (Urbana: University of Illinois Press, 1989); W. J. Rorabaugh, *The Craft Apprentice: From Franklin to the Machine Age in America* (New York: Oxford University Press, 1986).

18. The Davis-Bacon Act of 1931 is a U.S. federal law which established the requirement for paying prevailing wages on public works projects. All federal government construction contracts, and most contracts for federally assisted construction over $2,000, must include provisions for paying workers on site no less than the locally prevailing wages and benefits paid on similar projects; for a historical overview of prevailing wage policy in the construction industry, see Hamid Azari-Rad, Peter Philips, and Mark J. Prus (eds.), *The Economics of Prevailing Wage Laws* (Burlington, VT: Ashgate, 2005).

19. For a comprehensive discussion of the origins of the multiple gender and race-driven meanings of "master" craftsmen in the building and trowel trades in apprenticeship, slavery, and the American family, see Rorabaugh, *The Craft Apprentice*. See also Padavic and Reskin, *Women and Men at Work*, 2nd ed., 23.

20. A soldering iron is a gun-like device for applying heat to melt solder for soldering two metal parts together. For electrical work, wires are usually soldered to printed circuit boards, other wires, or small terminals. In the early part of the twentieth century, and in breach of collective bargaining agreements with the Local, employers of movie theaters in Manhattan hired nonunion workers who would pose as professionals to get past union lines. According to an interview with Harry Van Arsdale, these professionals would wear suits and carry a briefcase to the job site out of which they would pull their soldering irons. For more on the life and times of Harry Van Arsdale Jr., see "Harry Van Arsdale, Jr.: 1905–1986," *The Electrical Union World* 46, no. 4 (March 7, 1986): 1, 6–7.

21. The Institute for Women and Work (IWW) cosponsored the first unionized tradeswomen's meeting of Tradeswomen Now and Tomorrow (TNT), a national organization of tradeswomen from across a broad spectrum of building trades work on April 1, 2001, at New York University's Wagner Labor Archives and at the District Council of Carpenters in New York City.

22. See Appendix D: Overview of the Construction Industry for a comparison of tradesmen's wages in various states throughout the United States. New York and New

Jersey are two of the highest ranking states in terms of high mean annual salaries of electricians. Reportedly, Local 3 electricians represent some of the highest hourly wage workers in the world.

23. For the specific divisions in Local 3, see Appendix D.

24. Depending on the influence of certain social clubs within the Local, members can also access information regarding potentially privileged job sites which are characterized by access to overtime pay, light duty in job assignments, and a greater longevity of the project, as opposed to short-term projects, jobs at pollution plants, potentially hazardous work (e.g., removing asbestos), and jobs at other work sites in New York that are dangerous.

25. See George E. Marcus, "Contemporary Problems of Ethnography in the Modern World System," in James Clifford and George E. Marcus, eds., *Writing Culture: The Poetics and Politics of Ethnography* (Berkeley: University of California Press, 1986), 165–193.

26. James Clifford, "Partial Truths," in Clifford and Marcus, eds., *Writing Culture*, 4.

27. An open shop in U.S. labor relations is a place of employment at which one is not required to join or financially support a union as a term and condition of hiring or continued employment; for example, open shops are required by law in several "right to work" states. In contrast, a closed shop is one in which an individual is required to become a member of a union prior to being employed or as a term or condition of continued employment.

28. Molly Martin, *Hard-Hatted Women: Stories of Struggle and Success in the Trades* (Seattle: Seal Press, 1988); Susan Eisenberg, *We'll Call You When We Need You* (New York: ILR Press, 1998); Jean Reith Schroedel, *Alone in a Crowd, Women in the Trades Tell Their Stories* (Philadelphia: Temple University Press, 1985); Kris: *Why Working Class Men Put Themselves and the Labor Movement In Harm's Way* (Ithaca: Cornell University Press, 2006), *Working Construction: Why Working-class White Men Put Themselves and the Labor Movement in Harm's Way* (Ithaca, NY: Cornell University Press, 2006).

29. See Clifford and Marcus, eds., *Writing Culture*. See also William Roseberry, *Anthropologies and Histories: Essays in Culture, History, and Political Economy* (New Brunswick, NJ: Rutgers University Press, 1994), 145–197.

30. W. J. Rorabaugh, *The Craft Apprentice: From Franklin to the Machine Age in America* (New York: Oxford University Press, 1988).

31. Sean Wilentz, *Chants Democratic: New York City & The Rise of the American Working Class, 1788–1850* (New York: Oxford University Press, 1984).

32. David Montgomery, *Citizen Worker: The Experience of Workers in the United States with Democracy and the Free Market System During the Nineteenth Century* (Cambridge: Cambridge University Press, 1993).

33. Eric Hobsbawm, *Workers: Worlds of Labor* (New York: Pantheon Books, 1984), especially Chapter 5, "The Transformation of Labour Rituals," 66–82.

34. Patricia A. Cooper, *Once A Cigar Maker: Men, Women, and Work Culture in American Cigar Factories, 1900–1919* (Chicago: University of Illinois Press, 1987).

35. Ileen A. De Vault, United Apart: Gender and the Rise of Craft Unionism (New York: Cornell University Press), 2004.

36. Alice Kessler-Harris, *A Woman's Wage: Historical Meanings and Social Consequences* (Lexington: University Press of Kentucky, 1990).

37. Alice Kessler-Harris, *In Pursuit of Equity: Women, Men and the Quest for Economic Citizenship in 20th Century America* (Oxford: Oxford University Press, 2003).

38. Nancy MacLean, *Freedom Is Not Enough: The Opening of the American Workplace* (Cambridge, MA: Harvard University Press, 2006).

39. Cynthia Cockburn, *Gender and Technology in the Making* (London: Thousand Oaks, 1993); also see Cockburn, *Brothers: Male Dominance and Technological Change* (London: Pluto Press, 1983); also for a discussion of the types of organizational

arrangements and networks that restrict women's inclusion in male-dominated work settings, see Cockburn, *In the Way of Women: Men's Resistance to Sex Equality in Organizations* (Ithaca, NY: Cornell University Press, 1991). The term "separate spheres" is defined in the feminist literature as a set of beliefs originating among the British upper-middle classes which called for the separation of work and family life. According to Clawson, this ideology promoted the idea that a woman's place was in the home and a man's place was in the market conducting commerce. See also Padavic and Reskin, *Women and Men at Work*, 2nd ed., 42–45, for a discussion of how the ideology of separate spheres has led to sex segregation in the workplace.

40. Mary Ann Clawson, *Constructing Brotherhood: Class, Gender and Fraternalism* (Princeton: Princeton University Press, 1989).

41. Ibid.

CHAPTER 1. BROTHERHOOD

1. Mary Cummings, "Bayberry Land," available at www.hamptons.com/hamptons-article-magazine-568.htm, accessed September 10, 2004.

2. Ibid., p. 20.

3. "The History of Local 3 As I Remember It," May 17, 1976, Joint Industry Board of the Electrical Industry, Harry Van Arsdale Jr. Library at Empire State College (SUNY). This is an unpublished three-page typewritten memoir of events surrounding the Local's ascent to power in New York City in the early part of the twentieth century by an anonymous pioneer electrician.

4. Francine Moccio, interview with Harry Van Arsdale Jr., Electrical Apprenticeship Program, Charles Evans Hughes High School, September 18, 1985.

5. Francine Moccio, Field Notes, Bayberry Land Conference, Southampton, Long Island, July 3, 1992.

6. Ibid.

7. Alice Kessler-Harris, *A Woman's Wage: Historical Meanings and Social Consequences* (Lexington: University Press of Kentucky, 1990), 20.

8. Mary Ann Clawson, *Constructing Brotherhood*, 5–7.

9. Quoted in Clawson, *Constructing Brotherhood*, 5.

10. Grace Palladino, *Skilled Hands, Strong Spirits* (Ithaca: Cornell University Press, 2005), 55.

11. Grace Palladino, Peter J. Albert, and Stuart Kaufman, eds., *The Samuel Gompers Papers, Vol. 6: The American Federation of Labor and the Rise of Progressivism, 1902–6* (Urbana: University of Illinois Press, 1997).

12. In Palladino, *Skilled Hands*, 27.

13. Ibid.

14. Clawson, *Constructing Brotherhood*. See also Eric Hobsbawm, *Workers: Worlds of Labor* (New York: Pantheon Books, 1984), 69.

15. Sean Wilentz, *Chants Democratic*, 387. See also Rorabaugh, *The Craft Apprentice*. For a discussion of fraternalism and proto–trade unionism in London, see Hobsbawm, *Workers*, 68.

16. Clawson, *Constructing Brotherhood*, 58.

17. Montgomery, *The Fall of the House of Labor*, 182.

18. Ibid.

19. The function of secrecy among workers vis-à-vis employers is different from its function in relation to less skilled and immigrant workers. The former is defensive and protective; the latter is exclusionary.

20. Ronald W. Schatz, *The Electrical Workers: A History of Labor at General Electric and Westinghouse 1923–60* (Urbana: University of Illinois Press, 1983), 30–33, 120. According to Schatz, the electrical industry at this time consisted of 'male-typed'

craft work such as sheet metal workers, and 'female-typed' manufacturing jobs such as coil winders and assembly-line production in electrical appliance plants that manufactured cheap consumer products such as light bulbs and lamp shades. Female coil winders were known for their speed, attention to detail, nimble fingers, and ability to concentrate. Men were segregated into machine-operative jobs which demanded muscular strength, calculation, and training.

21. Palladino, *Skilled Hands*, 5–7, 16–21, 27, 199–201. See also Montgomery, *The Fall of the House of Labor*, 3–8, 15–17. Some employers sought to make an alliance with the building trades associations in order to keep the industrial peace as a result of the chaos and violence occurring on construction work sites in protest of employer exploitation of workers.

22. Palladino, *Skilled Hands*, 27.

23. Ibid.

24. Grabelsky, "Short Circuit."

25. Moccio, interview with Harry Van Arsdale Jr.

26. "Beginning of Brotherhood," in "The History and Structure of the International Brotherhood of Electrical Workers, Part 1," in *Electrical Union World*, 1986. See also "Labor History: Forming the International Brotherhood of Electrical Workers, Part II," *Electrical Union World*, January 5, 1986.

27. Michael A. Mulcaire, *The International Brotherhood of Electrical Workers: A Study in Trade Union Structure* (Washington, DC: IBEW, 1923), 8. See also Grace Palladino, *Dreams of Dignity, Workers of Vision: The International Brotherhood of Electrical Workers* (Washington, DC: IBEW, 1991). See also "Forging the Brotherhood, Part 1," *Electrical Union World*, January 15, 1986.

28. *Journal of Electrical Workers*, 41, no. 4 (April 1942).

29. Ibid.

30. Schatz, *The Electrical Workers*, 28, 29, 30.

31. Ilene A. DeVault, "Strong Hands to Aid the Weak: Gender in AFL Craft Unions, 1887–1894" (unpublished paper, 1990) in possession of the author.

32. According to Schatz (*The Electrical Workers*, 30), 91% of jobs in electrical manufacturing in 1939 were held by female operatives. Women were underrepresented among the leadership of their labor organization in the electrical industry and were segregated into 'female-typed' work. Women worked in close proximity to one another and developed close personal friendships that enabled them to develop a collective conscious to confront bosses. Consequently, according to Schatz (p. 33), "this also presented a paradox: women workers who ordinarily manifested little interest in unions proved to be the most tenacious fighters in shop-floor disputes and strikes."

33. C. C. Van Inwegen, Letter to Editor, *Electrical Worker* (March 1896), quoted in DeVault, "Strong Hands."

34. Philip S. Foner, *Women and the American Trade Union Movement: From Colonial Times to the Eve of World War I* (New York: The Free Press, 1979), 252–254.

35. Report #77 Seattle, *Electrical Worker* (August 1990), quoted in DeVault, "Strong Hands."

36. DeVault, "Strong Hands."

37. "Labor History: Forming the International Brotherhood of Electrical Workers, Part I," *Electrical Union World*, December 5, 1985; see also Patricia A. Cooper, *Once a Cigar Maker: Men, Women and Work Culture in American Cigar Factories, 1900–1919* (Urbana: University of Illinois Press, 1989). Cooper makes the point that this was also the case for Samuel Gompers' Cigar Makers International Union (CMIU). Gompers later became the leading figure in the AFL.

38. At this time it was not at all common or even possible for a woman factory worker to get a job as a construction electrician. At the onset, the creation of separate

locals for women workers was not necessarily intended as a device to exclude women from electrical apprenticeships, but a recognition that factory work and construction work were different enough to require separate (but unequal) jurisdictions. The manufacturing division of IBEW Local 3 resembles the early IBEW which was primarily an industrial union of electrical manufacturing workers.

39. W. J. Rorabaugh, *The Craft Apprentice*, 197. For a discussion of women's entry into other craft trades as less skilled workers and 'helpers' to apprentices in the tailor trade and union, see Ilene A. DeVault, *United Apart: Gender and the Rise of Craft Unionism* (Ithaca: Cornell University Press, 2004), 93.

40. *Union World*, January 15, 1986. See also Charles J. Madden Memorandum, Electrical Union No. 5468, Albany, New York, February 17, 1892; Memorandum, Electrical Union No. 5468, Brooklyn, New York, February 16, 1892; Address to Firms and Contractors in the Electrical Trade, Electrical Union No. 5468, New York, January 1, 1892.

41. Moccio, interview with Harry Van Arsdale Jr. See also Palladino, *Skilled Hands*, 161.

42. Miriam Frank, "The History of Local 3" (unpublished paper), Joint Industry Board of the Electrical Industry Archives, Flushing, New York, 1985, 1. See also Maurice Neufield, *Day In and Day Out With Local 3* (Ithaca: Cornell University ILR Press, 1955), 1.

43. "An Address to Firms and Contractors in the Electrical Industry," Executive Committee of Electrical Union No. 5468 to the Edison Illuminating Company, New York, January 1, 1892.

44. Grabelsky, "Short Circuit."

45. Neufeld, *Day In and Day Out*, 2.

46. In the early part of the twentieth century, when employers recognized the formidable power of the building trades unions and their workers to disrupt construction projects, employers agreed to work in cooperation with unionists throughout the year to avoid lockouts or strikes and to address possible problems and grievances that could delay projects and escalate building costs.

47. Moccio, interview with Harry Van Arsdale Jr.

48. Ibid.

49. Alice H. Cook, "Local 300: Guided Democracy," in *Union Democracy: Practice and Ideal, An Analysis of Four Large Local Unions* (Ithaca, NY: Cornell University Press, 1963), 113–143. For a description of clubs as networks based on kinship and politics, see "Assembly Districts Played A Major Role," *Electrical Union World*, November 15, 1961.

50. Moccio, interview with Harry Van Arsdale Jr. I also attended the Honor Scroll Night where several pioneers of the Local recounted their memories of the evolution of the Van Arsdale dynasty in Local 3.

51. "The History of Local 3 As I Remember It."

52. Moccio, interview with Harry Van Arsdale Jr.

53. Ibid.

54. "The History of Local 3 As I Remember It."

55. Palladino, *Skilled Hands*.

56. Moccio, interview with Harry Van Arsdale Jr.

57. Lincoln Steffens was the editor of a radical political magazine entitled American Magazine, and together with other radical investigative reporters of his time, like Ida Tarbell, he covered the Mexican Revolution and backed the Mexican rebels in his reports. He also visited the Soviet Union in 1919 and is known for his statement: "I have seen the future and it works." Steffens wrote his autobiography in 1931, when Van Arsdale was struggling for power over Local 3. By the time of his death in 1936, Steffens was writing about his disillusionment with the turn of events in the Soviet Union.

58. Moccio, interview with Harry Van Arsdale Jr., Charles Evan Hughes High School, Empire State College Labor Studies Program, 1985.

59. Ibid.

60. Ibid.

61. "History of the International Brotherhood, Part II," *Electrical Union World*, January 5, 1986.

62. Ibid.

63. *Electrical Union World*, 1943, 3. Joint Industry Board of the Electrical Industry Archives in Flushing Queens, New York. Describes, in part, the function of the fraternal clubs.

64. Moccio, interview with Harry Van Arsdale Jr.

65. Warren Moscow, "The Early History of Local 3" (unpublished manuscript, 1970), Joint Industry Board of the Electrical Industry Archives, Flushing, New York. The history of Bayberry Land can also be found in "Allied Club Hosts Bayberry Land Seminar," *The Electrical Union World News*, November 11, 1986.

66. Francine Moccio interview with Edward J. Cleary, former President of the New York State AFL-CIO, former President of IBEW, Local 3, May 27, 1993.

67. "Electrical Welfare Club Marks 61 Years of Progress," *Electrical Union World*, November 15, 1984. See also Jerry Finkel, "Electrical Welfare Club Enters 65th Year," *Electrical Union World*, February 22, 1988.

68. Reference here on the anniversary celebration of the *Electrical Union World* articles on the Electrical Welfare Club

69. "Prejudice Is Expensive," *Electrical Union World*, January 1948.

70. "Iglesias Society Marches in Puerto Rican Day Parade," *Electrical Union World*, January 1948. See also "Thanks to People of the Common Wealth of Puerto Rico," *Electrical Union World*, September 15, 1958.

71. "Members Tell of Ten-Day Visit to Virgin Islands and Jamaica," *Electrical Union World*, New York City, 1965.

72. "The Black Club Marks Silver Anniversary," *Electrical Union World*, December 15, 1983.

73. "Famed Inventor's Grandson Attends Latimer Dance," *Electrical Union World*, February 1, 1962.

74. Francine Moccio, anonymous focus group interview with black male electricians, New York City, December 20, 2006.

75. Moccio, interview with Harry Van Arsdale Jr.

76. *Allied Union News*, Joint Industry Board Archives of the Electrical Industry, Flushing Queens; Moccio interview with Harry Van Arsdale Jr., September 18, 1985. This conclusion is also drawn from a careful review of the fraternal club structure and descriptions in Local 3 brochures and other public relations materials, as well as interviews with club founders and officers. See also a special edition of the club's brochure entitled, "Everyday Religion in A Workaday World," Chapter #80 Community Breakfast Brochure, December 4, 1988, wherein the specific goals of the Catholic Council are enumerated, Joint Industry Board of the Electrical Industry Archives, Flushing, New York.

77. "Electrical Square Club Installation," *Electrical Union World*, March 1, 1956.

78. Moccio, interview with Edward J. Cleary, May 27, 1993.

79. "New York City Cracks Down on Unlicensed Electrical Work," *Electrical Union World*, March 17, 1988. As women increasingly showed interest in electrical building trades' work, the union experienced a tremendous growth of undocumented workers who were willing to work for wages below the standard set by federal law of the Davis-Bacon Act. See also "Protecting the Davis-Bacon Act," *Electrical Union World*, October 15, 1987; "Union Busters Lose Big One," *Electrical Union World*, May 1, 1986. The

Reagan years proved detrimental to Local 3's fight for enforcement of prevailing wage laws, see "White House Continues Attack on Davis-Bacon Act," *Electrical Union World*, May 1, 1986.

80. Reportedly, electricians are still the primary breadwinners in their families. If their wives work, it is usually in part-time jobs and in full-time traditionally female occupations such as nursing, teaching, and administrative support.

81. *Electrical Union World*, "Labor, Electrical Industry Leaders Hail Apprentice Training Instructors," February 10, 1970.

82. Moccio, interview with Harry Van Arsdale Jr., September 18, 1985.

83. Francine Moccio, interview with anonymous Levittown manufacturing worker, Williamsburg, Brooklyn, June 10, 1998.

84. Moccio, interview with Edward J. Cleary, May 27, 1993.

CHAPTER 2. A CLOSER LOOK AT LOCAL 3

1. Moccio, Interview with Harry Van Arsdale Jr., September 18, 1985; for an in-depth and detailed history of the life and times of Harry Van Arsdale Jr., see Gene Ruffini, Harry Van Arsdale Jr., *Labor's Champion* (Armonk, NY: M.E. Sharpe, 2002).

2. Ibid.

3. Ibid.

4. Palladino, *Skilled Hands*.

5. Neufeld, *Day In and Day Out*. This is not to imply that Harry Van Arsdale was a communist or had formal links to the U.S. Communist Party (CP). In fact, Van Arsdale, a very close friend and ally of David Rockefeller, disavowed any such links to the CP by becoming a card carrying member of the John Birch Society which was founded in 1958 during the anti-communist hysteria of the McCarthy Era for the purpose of fighting so-called enlightened socialists and other benevolent dictators. See also "Our Party versus Communism," *Electrical Union World*, 1961. Van Arsdale also made a point of differentiating the Brotherhood Party formed by the Local to help elect John Lindsey as mayor of New York City from the activities of the U.S. Communist Party.

6. Cook, "Local 300," 113–143.

7. Under the leadership and guidance of Van Arsdale and Jeremiah P. Sullivan for Local 3, employers such as A. Lincoln Bush, E. A. Kahn, and Harry Fischbach brokered an agreement for a "Voluntary Code of Fair Competition for the Electrical Contracting Industry of New York," which was signed on January 1, 1939. As part of the agreement, a provision was also made to establish the Joint Pension Plan of the Electrical Industry and later, in 1943, based on an employer-union leadership, the JIB was founded. "The Chairman's Statement," in *History and Organization of the Joint Industry Board of the Electrical Industry, 50 Years of Labor-Management Relations 1943–1993*, ed. Jerry Finkel (publication of the Joint Industry Board of the Electrical Industry, 1994), 5; booklet in possession of the author.

8. "The Beginning," in *History and Organization*, 7–10

9. Neufeld, *Day In and Day Out*, 1. It is also important to note that at the beginning of the JIB's founding, there were three employers who served as unpaid part-time chairmen through various terms in the years 1943 to 1966. Another example of father to son unionism that was a part of Local 3 culture is Armand D'Angelo, who was elected full-time Chairman of the Board in 1966, retired in 1982, and was succeeded by his son, Joseph D'Angelo, who served as Chair until 1986.

10. For detailed information on the broad array of benefits the JIB afford electrician journeymen, see the JIB web site at www.jibei.org/about.asp

11. Neufield, *Day In and Day Out*, 12.

12. Moccio, interview with Edward J. Cleary, May 27, 1993.

13. Ibid. See also Cook, "Local 300," 113–143.

14. Moccio, interview with Harry Van Arsdale Jr., September 18, 1985.

15. Ibid.

16. Moccio, interview with Edward J. Cleary, May 27, 1993.

17. Roger Waldinger, *Still the Promised City? African-Americans and New Immigrants in Postindustrial New York City* (Cambridge: Harvard University Press, 1996), 101–103.

18. Clawson, *Constructing Brotherhood*, 209–220. Clawson discusses the transformative nature of fraternal societies and workmen's associations in depth over the late nineteenth and twentieth centuries.

19. Cook, "Local 300," 120–126.

20. Moccio, interview with Harry Van Arsdale Jr., September 18, 1985. Some present-day construction unions in Australia still have these types of saloons which cater to construction workers with back rooms reserved for family gatherings.

21. Moccio, interview with Harry Van Arsdale Jr., September 18, 1985. In working-class communities, the barroom or the saloon often had a meeting hall in a back room. Usually, neighborhood people held wedding receptions or family parties. It was in these community saloons where men gathered after work.

22. Moccio, interview with Harry Van Arsdale Jr.

23. *History and Organization of the Joint Industry Board*.

24. Ibid., 7.

25. Ibid., 10, 11.

26. Ibid.; Cook, "Local 300," 23.

27. Warren Moscow, "Early History of Local 3" (unpublished manuscript, 1970), 50–53.

28. Neufield, *Day In and Day Out*, 2.

29. Moccio, interview with Harry Van Arsdale Jr., September 18, 1985.

30. Francine Moccio, interview with Lafayette "Buddy" Jackson, former director of the apprentice program, Joint Industry Board of the Electrical Industry, Flushing Queens, March 15, 1990 and April 2, 1992.

31. Noam Chomsky, "Interpreting the World," in *The Chomsky Reader* (New York: Pantheon Books, 1987), 81.

32. This division is comprised of immigrant workers who do not speak English, they work with dangerous substances doing work like asbestos removal, or working in polluted plants in the outer boroughs.

33. "Labor History, Part I," *Electrical Union World*, December 5, 1985.

34. This is an expression commonly used in the trade.

35. These observations result from an examination of past issues of *Electrical Union World (EUW)*, formerly stored at the Harry Van Arsdale Jr. School for Labor Studies (SUNY) in Manhattan and can now be found at the Joint Industry Board of the Electrical Industry Archives in Flushing Queens. I examined issues from the first issue in 1938 until 2005; I also interviewed one of the former editors of the *EUW*.

36. Francine Moccio, interview with former President of Local 3, at Local 3 Headquarters in Flushing Queens, October 16, 1989.

37. See Dorothy S. Cobble, *The Other Women's Movement: Workplace Justice and Social Rights in Modern America* (Princeton: Princeton University Press, 2004), 69–90, for an important discussion of different class interests in the development of women's rights public policy, such as the Equal Rights Amendment (ERA), and such court cases as *International Union, United Automobile, Aerospace & Agricultural Implement Workers of America, UAW, Et. Al. v. Johnson Controls, Inc.* Certiorati to the United States Court of Appeals for the Seventh Circuit, No. 89-1215. Argued October 10, 1990, Decided March 20, 1991. See also Alice Kessler-Harris, *In Pursuit of Equity: Women, Men, and the Quest for Economic Citizenship in 20th-Century America* (New York: Oxford Uni-

versity Press, 2001), 207. Kessler-Harris presents an account of the influence of unions on representations of gender equity in twentieth-century public policy.

38. Moccio, interview with Evan Ruderman, June 1, 1989.

39. Rorabaugh, *The Craft Apprentice*, 4.

40. Ibid.

41. Ibid.

42. Palladino, *Skilled Hands*, 128.

43. The JIB issues "work tickets" for workers assigned to job projects in New York City.

44. Montgomery, *The Fall of the House of Labor*, 9–27, 296. Nonunion work in the construction craft trades has grown precipitously over the last sixty years; the hiring of immigrant non-English speaking and undocumented day laborers has reached an alarming 30 percent in New York City building trades; thereby reducing both the economic and political power of the building trades unions.

45. MacLean, *Freedom Is Not Enough*, 96.

46. "The Industry Capitalism Forgot" *Fortune Magazine*, August 1947.

47. Reportedly some members of the Teamsters union receive financial kickbacks from contractors and workers for placing minorities on construction sites. These mafia-controlled "minority coalitions" have reportedly undermined the Local's efforts at providing contractors with apprentices from their Division A program. The Teamsters have also challenged the hegemony of Local 3's apprenticeship program in the industry. The minority coalitions, despite their name, are run by criminal gangs who use violence to break up construction sites at night when they are empty and then extort jobs for minority men who are seeking jobs in the industry. For example, a construction site in the city will be damaged by criminal gang members under cover of night. The next morning, mobsters will call the contractor and threaten additional destruction to the site if jobs are not provided and/or extortion money paid. Once employed by the contractor, the mafia extorts money from these minority men for the "employment service." The Local has brought the Teamsters to court for breaching agreements regarding the training of apprentices, and these battles are still ongoing.

48. Francine Moccio, interview with Harry Van Arsdale Jr., September 18, 1985.

49. Francine Moccio, interview with anonymous electrician journeyman, New York City, December 20, 2006.

50. Ibid.

51. Francine Moccio, interview with Hirsch Electric contractor, Jamaica, Queens, October 10, 1991.

52. For a discussion of the volatility of the construction industry and the demand for construction labor for the earlier part of the twentieth century, see Montgomery, *Citizen Worker*. For the economics of the construction industry in New York City post-World War II, see Roger Waldinger, *Still the Promised City?*

53. Francine Moccio, interview with Harry Van Arsdale Jr., September 18, 1985.

54. Ibid.

55. Waldinger, *Still the Promised City?*

56. *Apprenticeship on the Move: A System in Review: The Story of the Electrical Industry's Training Program*, booklet, publication of the Joint Industry Board of the Electrical Industry, 1976, 29.

57. MacLean, *Freedom Is Not Enough*.

58. *Apprenticeship on the Move*, 28.

59. Francine Moccio, interview with Lafayette "Buddy" Jackson, March 15, 1990 and April 2, 1992.

60. Ibid.

61. "Who Qualifies to Be An Apprentice?" in *Apprenticeship on the Move*, 30-37. This publication describes the qualifications and entrance tests of unionized apprentices.

But ultimately, according to the majority of informants, sponsorship by a family relation in Local 3 still has the most influence regarding the acceptance or rejection of an applicant for apprenticeship.

62. Agreement between Local Union 3 and the New York Electrical Contractors' Association, June 30, 2008.

63. There is a lack of data on turnover for women. However, we can say with certainty that in the 1970s, the small number of women apprentices was primarily white, and the share of black women was a meager 8.3% (see Appendix C, Tables 4a and 4b for actual numbers). After this time period, the representation drastically increased for black women as they were hired at a much higher rate than white women. For example, from 8.3% apprenticeships in 1970, as of 2000 black women comprised 42.2%. Therefore, the pool of black women and white women is close to 50/50 in the skilled building trades. Nevertheless, the overall integration of women is meager. From a very slow but steady increase in the 1970s and 1980s (24.8% and 24%, respectively), women increased only a scant 4.1% in the 1990s, one of the indications of a strong male backlash.

64. Francine Moccio, interview with Lafayette "Buddy" Jackson, March 15, 1990 and April 2, 1992.

65. Ibid.

66. Francine Moccio, interview with Harry Van Arsdale Jr., September 18, 1985.

67. *Our Industry's Mosaic: A Portrait of Minority Participation in the Electrical Industry*, booklet, publication of the Joint Industry Board of the Electrical Industry and the Lewis Howard Latimer Society of Local 3.

68. The Amber Light Society is a newly established all female construction electricians' club discussed in greater detail in Chapter 6. The women in the Amber Light Society have recently made a very strong effort to persuade the male leadership of the Local that it is time to have a female on the Local's influential Executive Board, but to no avail.

69. Francine Moccio, interview with Edward J. Cleary, May 27, 1993.

70. Francine Moccio, interview with Evan Ruderman.

71. Ibid.

72. Francine Moccio, anonymous interview with electrician journeywoman and member of the Amber Light Society, New York City, 2007.

73. Connie Fletcher, *Breaking and Entering* (New York: Pocket Books, 1995). See also Clara Bingham and Laura Leedy Gansler, *Class Action* (New York: Anchor Books, 2002), 100–111.

74. This is true as well for other building construction unions (and quite unusual for other industrial, service sector, and occupational unions) even for men. Grievances brought by women generally are not resolved and they cannot get remedy through the process. The Local is reluctant to follow up on a grievance against the employer because it fears contractors will hire nonunion labor or the contractor will perceive unionized labor as more costly. Straw bosses defend the right of male workers regardless of their behavior toward women. These "managers" tend to circle the wagons around male electricians and male contractors when it comes to defending male workers against charges of discrimination. Straw bosses believe that defending women workers will incur the wrath of members of their primarily male work crews for having "let a brother down." These crews can retaliate by slowing down production and impeding the overall progress of the project; thereby, jeopardizing the job of these quasi-managers.

75. Apprentice Orientation Kit: Your Estate, packet and booklet publication of the Joint Industry Board of the Electrical Industry. The packet and booklets contain the rights and duties of all apprentices as well as the benefits package and work rules of the trade.

76. "Rules of the Brotherhood of Electrical Workers," No. 3–Labor Assembly 5468, Knights of Labor, O'Donnell Printing Company, 6th Avenue, Section 2, 1897, 15, in Apprentice Orientation Kit: Your Estate.

77. Francine Moccio, interview with Lafayette "Buddy" Jackson, March 15, 1990 and April 2, 1992.

78. Apprentice Orientation Kit: Your Estate.

79. Ibid., 16.

80. Francine Moccio, interview with Lafayette "Buddy" Jackson, March 15, 1990 and April 2, 1992.

81. Furthermore, there are clubs for members of each major religion, ethnicity, and nationality represented in the union, as well as a club for women members, a club for retirees, and clubs for various staff positions within the union, such as the "Key" club which is a foremen's club.

82. Francine Moccio, interview with Harry Van Arsdale Jr., September 18, 1985; Francine Moccio, field notes from attendance at a meeting of the Bronx Acorn Club, March 1994.

83. This history of social clubs and borough clubs was also described in an event program brochure (accessed at HVA Jr. School for Labor Studies, Empire State College [SUNY]) for the club's golden anniversary party, held in Electchester, NY, and supplemented with an interview with Edward J. Cleary.

84. Harry Van Arsdale Jr.'s role in expanding the importance of clubs throughout his career cannot be overstated. By the end of his career he created clubs representing different ethnic and religious backgrounds, geographic residence, and even sports and entertainment.

85. Local 3 Social Clubs

CLUB	MEMBERSHIP
Allied Social Club	Workers who live in Queens
Amber Light Society	Female construction workers
Asian American Cultural Society	Asian workers
Bedsole Kingsborough Club	Workers who live in Brooklyn
Bronx Acorn Electric Club	Workers who live in the Bronx
Catholic Council	Catholic workers
Central Colony Club	Workers who live in Manhattan
Electrical Square Club	Masons
Electrical Unity Club	Shipyard Workers
Electrical Welfare Club	Jewish workers
Futurian Society of America	College Educated members
Greek Orthodox Council	Greek, religious workers
Keystone Club	Foremen
L.U. #3 Retirees Association	Retirees
Louis Howard Latimer Progressive Association	African American members
Motorcycle Club	Motorcycle Enthusiasts
St. George Association	Protestant workers
Santiago Iglesias Education Society	Hispanic workers
Ski Club	Ski enthusiasts
Spartan 419 Club	Manufacturing division workers
Sportsman Club of New York	Sports enthusiasts
Staten Island Electrical Club	Workers who live in Staten Island
Sword of Light Pipe Band	Pipe Band members
Westchester Mechanics Association	Workers who live in Westchester

86. Cynthia Harrison, *On Account of Sex: The Politics of Women's Issues 1945-1968* (Berkeley: University of California Press, 1988), 11.

87. Cook, "Local 300," 126.

88. Disciplinary actions are never taken by the Apprentice Advisory Association (AAA) members; however, the clubs are used as laboratories for observing ways in which apprentices adapt themselves to the political climate and status quo of the local's leadership. Apprentices who question policies or practices of the Local are quickly dismissed for future leadership roles. Values of the "unselfish" brotherhood are promoted in the AAA for example, personal dignity and cultivation of the individual. For the history of these values among journeymen, see Montgomery, *The Fall of the House of Labor*, 17.

89. Francine Moccio, anonymous interview with male journeyman, March 18, 2003.

90. Ibid.

91. Francine Moccio, interview with Harry Van Arsdale Jr., September 18, 1985.

92. Ibid.

93. *Electrical Union World*, "80 'A' Apprentices Receive Community Service Certificates," February 1, 1984.

94. Francine Moccio, interview with Harry Van Arsdale Jr., September 18, 1985.

95. Cook, "Local 300," 127.

96. Neufield, *Day In and Day Out*, 383–384.

97. Ibid, 384.

98. Cook, "Local 300," 128.

99. Francine Moccio, interview with former President of Asian-American Club, April 15, 2004.

100. Reportedly, there are now fledging South Asian clubs reflective of the most current ethnic layering in New York City.

101. Francine Moccio, interview with Evan Ruderman. For divisions among union women along the lines of class and race, see also Cobble, *The Other Women's Movement*, 7–8.

102. "Electchester May Be Completed By December," *Electrical Union World*, September 15, 1951.

103. Lester Velie, "The Union That Gives More to the Boss," *Reader's Digest* January 1956; "French Workers, Industrialists Visit Our Housing Project," *Electrical Union World*, June 1951; "O'Dwyer Congratulates Us on Housing Project," *Electrical Union World*, October 18, 1949; "A New Housing Partnership," *New York Times*, May 14, 1949.

104. David W. Chen, "Electchester Getting Less Electrical; Queens Co-op for Trade Workers Slowly Departs from Its Roots," *New York Times*, March 15, 2004, http://query.nytimes.com/gst/fullpage/html (accessed May 29, 2007). In 2004 the *New York Times* reported that resident electricians were treating Electchester more as a springboard to the suburbs. Increasing numbers of residents were falling behind on their payments, so Electchester officials were relying more on stringent credit checks to screen applicants.

105. "Electchester May Be Completed By December," *Electrical Union World*, September 15, 1951.

106. Chen, "Electchester Getting Less Electrical."

107. Francine Moccio, Field Notes, Flushing, New York, April 12, 1992.

108. Chen, "Electchester Getting Less Electrical."

109. Ibid.; see also "Our Pioneer Members Were There," *Electrical Union World*, September 15, 1961.

110. Moccio, interview with former president of Local 3.

111. "'A' Apprentices Elect New Advisory Board Officers," *Electrical Union World*; "Electrical Welfare Club Installs New Officers," *Electrical Union World*, March 23,

1983. "Business Manager Van Arsdale Installs Allied Club Officers," *Electrical Union World*, January 15, 1984.

112. Moccio, interview with former President of Local 3. "Meet Local 3's Apprentice Instructors," *Electrical Union World*; "Ladies Night Held By Bedsole Club and Catholic Council," *Electrical Union World*, April 7, 1986; "Latimer Group Makes First Scholarship Grant," *Electrical Union World*, October 1, 1961; "Local 3 Members Donate Toys," *Electrical Union World*, January 17, 1988.

113. Moccio, Field Notes, Bayberry Land Conference.

114. "Gov. Dewey Says Local 3's Housing Project Should Be 'Model' for U.S.," *Electrical Union World*, September 15, 1949. James Sigurd Lapham, "Indians . . . To Farms . . . To Golf Courses . . . To Today This Was Electchester," *Electrical Union World*, October 10, 1975.

115. "Gov. Dewey"; "How Truman Fought For Housing," *Electrical Union World*, September 1, 1949.

116. "A New Housing Partnership," *New York Times*, May 14, 1949; "Builders Start Work on New Community of 400 Houses in Westchester Country," *New York Times*, October 1950.

117. Francine Moccio, interview with former President of Local 3, Harry Van Arsdale Jr. School for Labor Studies (SUNY), Empire State College, New York, September 18, 1985.

118. "Are You Ready for the BIG Saturday Night?" *Electrical Union World*, September 1950; "Local 3 Honors Scroll Ladies," *Electrical Union World*, 1975; "Members' Wives and Children Enjoy Local 3's Outing at Steeplechase," *Electrical Union World*, July 15, 1950; "Local 3 Officials Installed," *Electrical Union World*, July 15, 1950; "Stork Visits Electchester First Time," *Electrical Union World*, June 1, 1952; "Ladies Night Held By Bedsole Club and Catholic Council," *Electrical Union World*, April 7, 1986.

119. "Our Party vs. Communism" *Electrical Union World*, September 15, 1961; "Party of the Bridge," *Electrical Union World*, September 15, 1961; "Mayor Wagner Tells Shop Stewards Labor Has Vital Stake in City's Election," *Electrical Union World*, September 1, 1961. In New York, election laws allow "fusion"—that is, candidates for any public office can run as the nominee of more than one political party. The votes candidates receive are tallied separately by party, and then combined. See also, Alyssa Katz, "The Power of Fusion Politics," *The Nation*, August 25, 2005 (on fusion politics in New York State as an electoral strategy).

120. "Brotherhood Party Rolls Along in High Gear," *Electrical Union World*, September 15, 1962.

121. "On the Line of Brotherhood Party," *Electrical Union World*, November 1, 1961; "Brotherhood Party Is Not a 'Class Party' in Attack on Red Ideology," *Electrical Union World*, October 6, 1961; "Central Labor Council Endorses Brotherhood Party," *Electrical Union World*, September 1, 1961; "Mayor Wagner Tells Shop Stewards Labor Has Vital Stake in City's Election," *Electrical Union World*, September 1, 1961. Martin Luther King Jr. was not the only civil rights leader the Local recruited. Julian Bond, the iconic American social activist and civil rights leader, was also invited by the Santiago Iglesias club (the Hispanic club) to speak at a 1973 Local 3 apprentice graduation; see "Santiago Iglesias Club Hosts Julian Bond," in the *Electrical Union World,* January 15, 1973.

122. "Van Arsdale Praises Rev. King for Leadership in Civil Rights Movement," *Electrical Union World*, April 1, 1964.

123. "Our Pioneer Members Were There," *Electrical Union World*, September 15, 1961.

124. "Van Arsdale Says: Our Brotherhood Party Is Here to Stay," *Electrical Union World*, November 16, 1961; "Van Arsdale Speech on Brotherhood Party Program," *Electrical Union World*, October 6, 1961.

125. While it is true that the AFL craft unions under Gompers did not usually participate in electoral politics—1912 has to be noted as the exception to this rule. New York labor history accounts of labor-based political parties, many with union support, abound in the literature. For examples of trade unionism and politics, see Wilentz, *Chants Democratic*, 386; Hobsbawm, Workers, 162, 198, 210–211, 289–290; Montgomery, *Citizen Worker*, 145–157.

126. "Van Arsdale Speech on Brotherhood Party Program," *Electrical Union World*, October 6, 1961.

CHAPTER 3. THE STRUGGLE TO BECOME ELECTRICIANS

1. Francine Moccio interview with Melinda Hernandez, electrician journeywoman, Bronx, June 19, 1989.

2. Alice Kessler-Harris, *Out to Work: A History of Wage Earning Women in the United States* (Oxford: Oxford University Press, 1982); Carol Chetkovich, *Real Heat* (New Brunswick: Rutgers University Press, 1997); Christine L. Williams, *Gender Differences At Work: Women and Men Nontraditional Occupations* (Berkeley: University of California Press, 1989); Cynthia Cockburn, *Machinery of Dominance: Women, Men and Technical Know-how* (London: Pluto Press, 1985); Isabella Stone, *Equal Opportunities in Local Authorities* (London: HMSO, 1988).

3. Ruth Milkman, *Gender at Work* (Urbana: University of Illinois Press, 1988).

4. Kessler-Harris, *A Woman's Wage*.

5. Ibid..

6. Milkman, *Gender at Work*.

7. Ibid.

8. DeVault, *United Apart*.

9. Ilene DeVault cites this fear among male UAW workers regarding women's unfair treatment and its eventual impact upon male membership. This level of consciousness has not yet reached male workers in the construction craft trades.

10. Milkman, *Gender at Work*.

11. Ibid. See also Bruce J. Dierenfield, "Conservative Outrage: The Defeat in 1966 of Representative Howard W. Smith of Virginia," *Virginia Magazine of History and Biography* (1981) 89, no. 2, 181–205.

12. Milkman, *Gender at Work*.

13. In Kessler-Harris, *In Pursuit of Equity*.

14. MacLean, *Freedom Is Not Enough*, 70.

15. Kessler-Harris, *In Pursuit of Equity*.

16. MacLean, *Freedom Is Not Enough*, 147, 269, 301–314.

17. President Kennedy's Executive Order 10925 of 1961 created the president's Equal Employment Opportunity Commission, and affirmative action required federal contractors to take positive steps to ensure equitable treatment of workers. Kessler-Harris, *In Pursuit of Equity*, 275.

18. Kessler-Harris, *In Pursuit of Equity*, 276.

19. "Overcoming Barriers to Successful Entry and Retention of Women in Traditionally Male Skilled Blue-Collar Trades in Wisconsin," in *Apprenticeship Research: Emerging Findings and Future Trends, Proceedings of a Conference on Apprenticeship Training*, April 30–May 1, 1980, Washington, DC, ed. Vernon M. Briggs, Jr. and Felician F. Foltman, p. 107.

20. MacLean, *Freedom Is Not Enough*, 147, 269, 301–314.

21. Maclean, *Freedom Is Not Enough*; Waldinger, *Still the Promised City?*

22. Roslyn D. Kane and Jill Miller, "Women and Apprenticeship: A Study of Programs Designed to Facilitate Women's Participation in the Skilled Trades." This study by Kane, of RJ Associates, Inc., and Miller, of the Institute for Women's Concerns, is

based on an 18-month study in 1977–78 sponsored by the U.S. Department of Labor, Division of Apprenticeship Research and Development, and published in *Apprenticeship Research: Emerging Findings and Future Trends, Proceedings of a Conference on Apprenticeship Training*, April 30-May1, 1980. The study surveyed women apprentices, interested applicants, employers, and unionists to determine why so few women applicants were being admitted to apprenticeship and to identify the differences between the successful and unsuccessful applicant.

23. In Kessler-Harris, *In Pursuit of Equity*, 271.

24. Ibid., 229. Pauli Murray believed that an Equal Rights Amendment would not be necessary if women were brought within the protective umbrella of the Fourteenth Amendment.

25. MacLean, *Freedom Is Not Enough*, 122.

26. In Kessler-Harris, *In Pursuit of Equity*, 279.

27. Ibid.

28. Betty Friedan shaped the mission of the NOW based on the model of the NAACP.

29. In Kessler-Harris, *In Pursuit of Equity*, 279.

30. Ibid.

31. Ibid.; see legal case: 467 F.2d 95, 4 Fair Empl.Prac.Cas. 1255, 5 Empl. Prac. Dec. P 7956, *Mrs. Lorena W. Weeks v. Southern Bell Telephone and Telegraph Company*.

32. For a more in-depth study and analysis of the social forces surrounding the passage of the 1972 Equal Opportunity Act (EEO), see Kessler-Harris, *In Pursuit of Equity*, 246–273; MacLean, *Freedom Is Not Enough*, 70–71; Cobble, *The Other Women's Movement*, 36.

33. For a more detailed description of the benefits of EEO laws on women and affirmative action in general, see MacLean, *Freedom Is Not Enough*, 101–111

34. Francine Moccio, interview with Betty Friedan, Washington, DC, August 13, 2003.

35. Heidi Hartmann, Capitalism and Patriarchy. An early but still very relevant article discusses how "men as men" have an interest in women's subordination both at home and at work and maintaining a sex-segregated labor market. Women as a class in the low-wage labor market are kept subordinate to men as a class.

36. David M. Gordon, Richard Edwards, and Michael Reich, eds., *Segmented Work, Divided Workers: The Historical Transformation of Labor in the United States* (New York: Cambridge University Press, 1982).

37. MacLean, *Freedom Is Not Enough*, 266. MacLean reports that although women failed to make progress in the building trades in the 1990s and early part of the twenty-first century, their numbers in the trades did improve in the 1980s and shot up to 105,000 tradeswomen nationally by 1983—still only 2 percent of the industry's labor force—but a marked improvement nevertheless. MacLean cites the following for this data: *Rocky Mountain News*, March 28, box 33. Eisenberg, Susan. "We'll Call You If We Need You: Experiences on Women Working Construction," *The Journal of American History* 86, no. 1 (1999).

38. Diana Pearce, "The Feminization of Poverty: Women, Work and Welfare" *Urban and Social Change Review* 11 (1978): 28–36; Erin L. Kelly and Sara S. McLanahan, "The Feminization of Poverty: Past and Future," in Janet Saltzman Chafetz, ed., *Handbook of the Sociology of Gender*, 127–145 (Boston: Springer, 2006).

39. MacLean, *Freedom Is Not Enough*, 260.

40. *Advocates for Women et al. v. Ussery et. al.*, Civil Action No. 76-527, U.S. District Court for the District of Columbia (filed 13 April 1976) (later *Women Working in Construction v. Marshall*).

41. MacLean, *Freedom Is Not Enough*, 270. In April 1978 President Jimmy Carter issued Revised Order No. 4, containing affirmative action regulations setting separate

and specific goals for hiring women in the trades apart from those already established for minority men; for further description on the activist push for Revised Order No. 4, see MacLean, *Freedom Is Not Enough*, 269–279.

42. National Commission on Working Women of Wider Opportunities for Women, Women and Nontraditional Work; Office of the Secretary, Women's Bureau, U.S. Department of Labor, *Sources of Assistance for Recruiting Women for Apprenticeship: Programs and Skilled Nontraditional Blue-Collar Work* (July 1978).

43. Nontraditional Occupations Program for Women, U.S. Department of Labor, Office of the Secretary Women's Bureau, *Women in Nontraditional Jobs: A Program Model* (1978).

44. Walter Ruby, "Contractors Make it Tough for Women," *In These Times* (September 9–15, 1981).

45. Ibid.

46. *National Directory of Women's Employment Programs: Who They Are, What They Do* (Washington, DC: WOW, 1979); "Conference Agenda," *Women's Work Force Report* (1979), 4, box 2, WOW Papers.

47. North East Women in Transportation, *A Brief History of What ISTEA Means for Women*, Vermont.

48. Moccio, interview with Melinda Hernandez.

49. New York City Commission on Human Rights, *Hearings on Discrimination in the Construction Industry*, 1993.

50. Francine Moccio, interview with Joyce Hartwell, August 29, 1989.

51. MacLean, *Freedom Is Not Enough*, 267.

52. Ibid.

53. The New York City Commission on Human Rights and Office of Labor Services, *Hearings on Discrimination in the New York City Construction Industry*, March 1990: 254; see also All-Craft Center, "Evaluation of Training and Job Placement Program for Economically Disadvantaged Women in Non-Traditional (Skilled Trades) Work," [n.d.], esp. 3, box 1, UT Papers; Wider Opportunities for Women, "A Proposal for WOC-CNPR Women in Construction Compliance Monitoring Project" (Jan. 1980), box 30, CNPR.

54. Ibid.

55. Francine Moccio, interview with Joyce Hartwell, August 29, 1989.

56. Francine Moccio, interview with Lafayette "Buddy" Jackson, March 15, 1990 and April 2, 1992.

57. EEOC Reports Joint Apprentice Committee Statistics, Local 3, "Total Apprentices Accepted in Programs (1974-1988), A Breakdown According to Race and Sex (In Percentages)," obtained from the New York Apprenticeship Council, New York State Department of Labor Compliance Reports.

58. Across the majority of the building trades apprenticeships, including but not limited to the electrical apprenticeship, test scores are a combination of objective testing in algebra and other areas, as well as an "interview" at the Joint Industry Board. Unfortunately, the most subjective test, the interview, carries the greatest percentage (80 percent) in determining the acceptance or rejection of a candidate for apprenticeship.

59. Testimony of Joyce Hartwell at the New York City Commission on Human Rights, *Discrimination in New York City Construction Trades* (December 1993).

60. Francine Moccio, interview with Joyce Hartwell, August 29, 1989.

61. Testimony of Joyce Hartwell at the New York City Commission on Human Rights, Discrimination in New York City Construction Trades (December 1993).

62. Ibid.

63. Ibid.

64. Testimony of Susan Pardes at the New York City Commission on Human Rights, Discrimination in the New York City Construction Trades (December 1993).

65. New York State Apprenticeship Council, *Statistics of Women and Minorities*, 2006, New York State Department of Labor Compliance Reports On Entrance of Women and Minorities in Unionized Apprenticeship Programs.

66. Ibid.

67. David Dinkins and Dennis Deleon, *Building Barriers: A Report on Discrimination Against Women and People of Color in New York City's Construction Trades* (One Police Plaza, New York, 1993).

68. New York Department of State and City Statistics, *Bureau of Labor Statistics Geographic Profiles of Employment and Unemployment, 1984–1999*.

69. This information regarding the approximate number of female Division A members of Local was related to me in an interview with Cheryl Farrell, former president of The Amber Light Society, New York City, July 16, 2007.

70. New York State and New York City *2005 Census Data*; for the complete list of occupations, see www.labor.state.ny.us/workforceindustrydata/eeo_pr2asp?reg=nys& geog=01000036New%20

71. Francine Moccio, interview with Martha Baker. See also, Sharon Mastracci, *Breaking Out of the Pink Collar Ghetto* (New York: M.E. Sharpe, 2004).

72. Sharon H. Mastracci, "Persistent Problems Demand Persistent Solutions: Evaluating Policies to Mitigate Occupational Segregation by Gender," *Review of Radical Political Economics*, 37 Winter (2005), 23.

73. A total of 1,541 graduates of NEW's pre-apprenticeship program were analyzed in Francine Moccio and Marni Finkelstein, "Ethnicity of NEW Graduates: 1978 to 2007," *Interim Report on Women's Progress in NEW Pre-apprenticeship Training*, Cornell University, Institute for Women and Work (IWW), publication of the IWW and the Stephen P. Vladeck Memorial Fund (2008).

74. 2006 Equity Leadership Awards Luncheon: Non-Traditional Employment for Women, 2006.

75. Testimony of Susan Pardes at the New York City Human Rights Commission Hearings, 1990, p 36.

76. Ruth Milkman, "Women Workers and the Labor Movement in Hard Times: Comparing the 1930s with the 1980s" (1987); Cynthia Cockburn, *Brothers: Male Dominance and Technological Change* (London: Pluto Press, 1983); Cockburn, *Machinery of Dominance*; Cockburn, *Two-Track Training: Sex Inequalities and the Youth Training Scheme* (Basingstoke: Macmillan, 1987), Cockburn, *In the Way of Women: Men's Resistance to Sex Equality In Organizations* (New York: Industrial-Labor Relations Press, Cornell University, 1991); DeVault, *United Apart*.

77. Milkman, *Gender at Work*.

78. For example, women only comprise 1.5 percent of carpenters and nineteenths of a percent of pipe fitters, New York *Census Data*.

79. See Susan Faludi, *Stiffed: The Betrayal of the American Man,*(New York: William Morrow Co., 1999). On men's reactions to the feminist movement and self-definitions of masculinity, see also Michael Kimmel, *Manhood in America: A Cultural History* (New York: Oxford University Press, 2005).

80. Interview notes with Thomas Van Arsdale, former business manager and son of the Local's founder; see also DeVault, *United Apart*. DeVault writes about male workers' reservations about women entering local unions, as well as the fear on the part of the male union leadership that some day they might be "ruled by a woman." Moccio, interview with Thomas Van Arsdale, October 1990.

81. "Catholic Council Hosts 26th Annual Ladies Night," *Electrical Union World*, April 21, 1988; *Electrical Union World* 66, no 5, June 2005.

82. Martha Midgett, an African American Local 3 member from the Levittown Manufacturing Co. in Greenpoint, Brooklyn, became the first coordinator of the Women's Club of Local 3, which was organized in 1957 and was called the Women's Economy

Organization. In 1981 it was renamed the Women's Active Organizational Club and Midgett was on its executive board. Midgett is also a member of the Coalition of Labor Union Women. Other African American women figure prominently in leadership roles in the Local's manufacturing division, such as Vivian Merriwether, another EB member of the Women's Club, and Lola Barton, a member of the National Council of Negro Women, a trustee of the Educational and Cultural Fund of the Electrical Industry, a member of the NAACP, and former financial secretary of the Women's Club.

83. "Our Industry's Mosaic: A Portrait of Minority Participation in the Electrical Industry," 1995 publication of the Joint Industry Board of the Electrical Industry, in possession of the author.

84. MacLean, *Freedom Is Not Enough*, 9, cites Hill on the critical need to integrate black men into construction jobs, given before a congressional subcommittee in 1964 where he testifies about "injured manhood" among black men who are seen as "emasculated by racism and in need of a boost." Hill states a concern for black masculinity and "that these jobs [construction] were for men, they were 'male jobs' and 'manly jobs' with 'high status implications' especially important for Negro men."

85. Here it is interesting to note Shulamith Firestone's 1970 work "The Dialectic of Sex: The Case for Feminist Revolution," 118–141. Firestone argued that "racism is sexism extended" because, at least for Western cultures, it can only be understood in terms of the power hierarchies of the nuclear family. According to Firestone, race relations in America represent a macrocosm of the hierarchical relations within the nuclear family: The white man is father, the white woman wife-and-mother, her status dependent on his; the blacks, like children, are his property, their physical differentiation marking them the subservient class, in the same way that children easily become a servile class vis-à-vis adults. This power hierarchy creates the psychology of racism, just as in the nuclear family, it creates the psychology of sexism.

86. These alternate routes include organizing minorities on nonunion jobs through a process called "salting" on construction sites; that is, the Local sends a union member to work in a nonunion shop and he signs up nonunionized workers. In addition, in the 1970s and 1980s, the New York State Department of Labor ran an outreach program under State Labor Law 220 to encourage contractors to hire and train immigrant and minority workers other than those who had gone through the Division A apprenticeship program; however, State Law 220 was dismantled in court in 1987 by the New York State Department of Labor in response to pressure from electrical workers, see Waldinger, "Who Gets the Good Jobs?" in *Still the Promised City?* 193–194.

87. Moccio, interview with Melinda Hernandez.

88. Francine Moccio, interview with Laura Kelber, electrician journeywoman, New York City, June 5, 1990.

89. Francine Moccio, interview with Beth Schulman, electrician journeywoman, August 25, 1992.

90. MacLean, *Freedom Is Not Enough*.

91. Testimony of anonymous electrician journeywoman at the New York City Human Rights Commission Hearings on Discrimination in the Building Trades, 1990.

92. Although this is a process in the industry, recall men were settled already in other parts of the country with families and children in schools and they resented having to return to the New York area.

93. Francine Moccio, interview with Mary Ellen Boyd, transcript, March 1992.

94. Moccio, interview with Cynthia Long, May 5, 1992.

95. Francine Moccio, interview with Mary Ellen Boyd, transcript, March 1992.

96. Ibid.

97. Testimony, New York City Human Rights Commission Hearings, 1990, 45.

98. Moccio, interview with Evan Ruderman, June 1, 1989.

99. Testimony, New York City Human Rights Commission Hearings, 1990, 40.

100. It was at this time that I was asked to develop a training program and a video, and to conduct training for all officers, shop stewards and foremen in the Local.

101. Francine Moccio interview with Stan Brown, August 1990.

102. Francine Moccio interview with Harry Van Arsdale, Jr., Electrical Apprenticeship Program, Charles Evans Hughes High School, September 1985.

103. Moccio, interview with Evan Ruderman, June 1, 1989.

104. Francine Moccio, interview with former President of Local 3, Local 3 Headquarters, Flushing Queens, October 16, 1989.

105. Francine Moccio, interview with Tony Pecorello electrician journeyman, January 11, 1991.

106. Francine Moccio, interview with Stan Brown, August 1990.

107. Census Bureau Data, 2000 Report; also the December 31, 2008 New York State Apprenticeship Data on Females in the Construction Trades shows that women comprise nearly forty-seven percent of the New York State civilian labor force but are only seven percent of building trades apprentices, data obtained from the New York State Department of Labor, Office of the Training Coordinator.

108. Francine Moccio, interview with Tony Pecorello electrician journeyman, 1991.

109. Ibid.

110. Thomas Bailey and Roger Waldinger, "Primary, Secondary, and Enclave Labor Markets," American Sociological Review 56 (August 1991) 47; oral history interview with current Business Manager, Thomas Van Arsdale, at the IBEW convention in Atlantic City, New Jersey, May 1992.

111. Francine Moccio, interview with anonymous Puerto Rican journeyman, 2006.

112. Research on comparing regional differences in the acceptance of women as compared with minority men and women of color in the trades is one of the areas that needs further research.

113. The FLEAS are considered by tradesmen to be "a union within a union" and a nonsanctioned fraternity as per the IBEW's constitution. Respondents have related that members of the FLEAS are an active group of militant workers who work "underground" to try and shape policy within the union. Members of the FLEAS view themselves as more "worker" oriented and distinct from more employer-oriented business fraternities such as the free masons, a society that serves business interests. Scholars such as Waldinger and Bailey have suggested that placing restrictions on the use of travelers to fill expanding jobs in construction in peak periods may be a part of an effective overall strategy to expand the inclusion of an increased number of minorities in the trade.

CHAPTER 4. ON THE ELECTRICAL CONSTRUCTION WORK SITE

1. I had contact with and interviewed women electricians who worked in Division A in the years between 1978 and 2009. Unless otherwise noted all quotations from women electricians are from interviews with women who prefer to remain anonymous (names in quotation marks are pseudonyms): April 23, 2006, May 5, 23, 2006, March 12, 2008.

2. See Darla G. Hall, Patty J. Baxter and Jeanette Ticknow v. Gus Construction Co., Inc., 842 F.2d 1010. U.S. App. 1988; Eileen Lynch v. Charles H. Dean, Jr., et al. U.S. Dist. 1985l; E. Jones v. Intermountain Power Project, 794 F.2d 546. U.S. App. 1986; Marion J. Wells v. Colorado Dept. of Transportation. U.S. App.; U.T.B. United Third Bridge, Inc., v. IBEW NY, 512 F. Supp. 288. U.S. Dist. 1981; Annette Streeter and Ivette Ellis v. IBEW NY, 767 F. Supp. 520. U.S. Dist. 1991.

3. Francine Moccio, interview with the former director of the apprenticeship program, March 15, 1990 and April 2, 1992.

4. Francine Moccio, interview with Cheryl Smyler-George and Pat Sullivan, transcript, March 9, 1992; Francine Moccio interview with Ellen Kratenstein, fourth year

electrician apprentice, Charles Evans Hughes High School, June 22, 1989 and July 11, 1989; Moccio, interview with Evan Ruderman, June 1, 1989; Francine Moccio, interview with Cheryl Farrell, former President of The Amber Light Society, Cornell Conference Center, New York City, July 16, 2005.

5. Interview with the former director of All-Craft, one of the first not-for-profit organizations in New York City to integrate women into the construction trades, interview took place in September 2006 via telephone; notes in possession of the author.

6. Women electricians who participated in focus group sessions in 1990–98: Yvette Alvarez, Jackie Torres, Anna Hart, Judy Johanessen, Vera Boothe, Jo Neice Laramore, Stephanie Salvatore, Kathy Laury, Kathy Coley, Josephone Testa, Gladys Lopez, Yvette Ellis, Annette Streeter, and Cynthia Ellington, May 31, 1990; March 15, 1998. See also Francine Moccio, "Pioneers, Settlers or Passersby?: Women in the Skilled New York City Building Trades, An Evaluation of NEW's Apprentice Trades Program: 1980-2006," published report on NEW, Cornell University, Institute for Women and Work, 2006 (see Appendix B). Interviews with 100 female respondents who graduated from NEW's pre-apprenticeship program in a variety of skilled building trades, such as carpenters, electricians, plumbers, steamfitters, ironworkers, stone masons, and the like were conducted on May 22, 2003, June 30, 2006, and May 3, 2007; Cynthia Marano, *Wider Opportunities for Women Testimony*, The Commission of the Future of Worker-Management Relations, July 25, 1994; Francine Moccio, *Nontraditional Employment for Women Survey*, Cornell University, Institute for Women and Work, administered 1980–2007.

7. Irene Padavic and Barbara Reskin, "Men's Behavior and Women's Interest in Blue-Collar Jobs," *Social Problems* 37, no. 4 (November 1990). Padavic and Reskin found that among women who were heads of households and had children to support, hostile behavior they encountered in nontraditional jobs did not deter their interest or entry. For an analysis of the importance of race in women's choices for nontraditional jobs, see also Irene Padavic, "Attractiveness of Working in a Blue Collar Jobs for Black and White Women: Economic Need, Exposure, and Attitudes," *Social Science Quarterly* 72, no. 1 (1991). In this early study, Padavic finds that black women as compared with their white female counterparts are more willing to tolerate the inconveniences of nontraditional employment.

8. Carole G. Garrison, Nancy K. Grant, and Kenneth McCormick, "Perceived Utilization, Job Satisfaction and Advancement of Police Women," *Public Personnel Management* 19, no. 2 (Summer 1990); R. Stewart "Satisfaction of Women in Nontraditional Occupations," *Journal of Employment Counseling* 26 (1989); Sylvia A. Law, "Voices of Experience: New Responses to Gender Discourse," *Harvard Civil Liberties Law Review* 24, no. 1 (Winter 1989); Cynthia Marano, Wider Opportunities for Women, testimony to The Commission of the Future of Worker-Management Relations, July 25, 1994; Susan Eisenberg, "Women Hard Hats Speak Out," *Nation*, September 18, 1989; Natalie Solokoff, "What's Happening to Women's Employment: Issues for Women's Labor Struggles in the Late 1980s-1990s," in *Hidden Aspects of Women's Work* (New York: Praeger Publishers, 1987); J. Appier, *Policing Women: The Sexual Politics of Law Enforcement and the LAPD*. (Philadelphia: Temple University Press, 1998); American Community Survey 2006, www.census.gov/acs (accessed June 2008); Dan Bell, "Women Not Wanted: Female Construction Workers Face Chronic Unemployment and Daunting Odds. A New Mayoral Commission Will Have to Change the Face of an Industry," *City Limits*, May/June 2005, 12–15.

9. Moccio, interview with Evan Ruderman, June 1, 1989.

10. Francine Moccio, interview with Maura O'Grady, former electrical apprentice, World Trade Center, New York City, June 13, 1989.

11. Sara S. McLanahan and Erin L. Kelly, "The Feminization of Poverty: Past and Future," a publication of the Bendheim-Thoman Center for Research on Child Wellbeing, Department of Sociology, Princeton University, 2008.

12. American Community Survey 2006, www.census.gov/acs (accessed June 2008).

13. National Commission on Working Women of Wider Opportunities for Women, Women and Nontraditional Work; National Partnership for Women & Families, News Release. July 7, 2004.

For an international comparison of women's wages, see Elisabeth Hermann Frederiksen, "An Equilibrium Analysis of the Gender Wage Gap," Paper provided by Economic Policy Research Unit (EPRU), University of Copenhagen, Department of Economics, available at http://ideas.repec.org/s/kud/epruwp.html

14. For evidence of how sex segmentation in the labor market influences the narrowing or closing of the gender wage gap, see: Elisabeth Hermann Frederiksen, "An Equilibrium Analysis of the Gender Wage Gap," Paper provided by Economic Policy Research Unit (EPRU), University of Copenhagen.

15. Moccio, interview with Cynthia Long, May 5, 1992; Francine Moccio, interview with Bernard Rosenberg, President of the Electrical Welfare Club, Joint Industry Board Flushing Queens, March 20, 1990; Francine Moccio, interview with former President of the Electrical Welfare Club.

16. Moccio, interview with former business manager, October 1990.

17. Moccio, interview with electrical contractor, October 10, 1991; Moccio, interview with Richard Addeo, Manhattan, September 12, 1990; Moccio, interview with Susan Hayes and Lee Schrager, April 2, 1992.

18. Deck work is a large construction site often times employing hundreds of workers from across a broad spectrum of building trades work. These jobs, such as the rebuilding of downtown Manhattan and Harlem Hospital, among other sites in New York city, are federally subsidized, as such contractors are not required to hire a certain number of women but they are obligated to show the federal government that they have made a good faith effort to recruit women onto job sites.

20. Moccio, interview with Veronica Rose, New York City, April 8, 1992.

21. Ibid.

22. Francine Moccio, interview with Tony Pecorello, electrician journeyman, Empire State College, January 11, 1991.

23. Jeffrey Grabelsky, "Preserving Craft Pride: Autonomy and Control in the Construction Industry" (unpublished article); Grabelsky, "Short Circuit: The Light and Power Council Strike of 1913 and the Search for Effective Organizational Models in the International Brotherhood of Electrical Workers" (unpublished article); see also Cockburn, *Machinery of Dominance*; Montgomery, "The manager's brain under the workman's cap," in *Workers' Control in America: Studies in the History of Work, Technology and Labor Struggles* (New York: Cambridge University Press, 1979), 9–57; Dorrine K. Kondo, *Crafting Selves: Power, Gender, and Discourses of Identity in a Japanese Workplace* (Chicago: The University of Chicago Press, 1990), 300–308, a case study of masculine and feminine identity in a Japanese workplace; Palladino, *Skilled Hands*.

24. Moccio, interview with Harry Van Arsdale Jr., September 18, 1985.

25. A total of 1,234 workers were killed on the job in the private construction industry in 2007, with 10% (122) of them dying because of contact with electric current. The fatality rate for the private construction industry was 12.0 fatalities per 100,000 workers in 2007, fourth highest among industry sectors. The comparable figure for all workers was 4.1 (Bureau of Labor Statistics, "Occupational Injuries, Fatalities and Illnesses: Counts, Rates and Characteristics," 2006).

26. Agreement and Working Rules between New York Electrical Contractors Association, Inc., Association of Electrical Contractors, Inc., and Local Union No. 3 International Brotherhood of Electrical Workers, AFL CIO (June 13, 1986).

27. This machismo factor is also commonplace among male carpenters on work sites, see Kris Paap, *Working Construction*.

28. Francine Moccio, interview with Michael Wood, journeyman electrician, 1999.

29. This macho index, as well as the family wage argument, is emphasized by union leaders in the brotherhood as a bargaining strategy vis-à-vis contractors.

30. Moccio, interview with Evan Ruderman, June 1, 1989.

31. Ibid.

32. See Christine Williams for her description of war readiness preparedness among male recruits in the Marines.

33. Francine Moccio, interview with Harriet Shroup, electrician apprentice, Hunter College, April 5, 1989.

34. Moccio, interview with Evan Ruderman, June 1, 1989.

35. Authors such as Eisenberg, Martin, and Paap have observed this for construction; Chetkovich and Cockburn underscore this male bonding for firefighters and printers, especially when challenged with either dangerous work in the former case, or obsolesces in the latter.

36. Women in blue-collar communities, as noted by Cobble and Kessler-Harris among other scholars, were more concerned with issues having to do with the economic survival of their families, and local neighborhood issues, than with breaking into male-dominated occupations.

37. Carol Gilligan, *In a Different Voice: Psychological Theory and Women's Development* (Cambridge: Harvard University Press); Karen Sacks, *Sisters and Wives: The Past and Future of Sexual Equality* (Urbana: University of Illinois Press, 1982); Lillian B. Rubin, *Worlds of Pain* (New York: Basic Books, 1976).

38. Francine Moccio, interview with Emily Forman, electrician apprentice, West Bank Café, June 26, 1989.

39. Moccio, interview with Evan Ruderman, June 1, 1989.

40. Nancy Chodorow, "Family Structure and Feminine Personality," in *Women, Culture and Society*, 49-50; Andrew Tolson, *The Limits of Masculinity* (London: Tavistock Publications, 1977); Barbara Reskin and Patricia Roos, *Job Queues, Gender Queues* (Philadelphia: Temple University Press, 1990), 3–68. See also Carol Bacchi, *Same Difference: Feminism and Sexual Difference* (Sydney: Allen and Unwin, 1990); Chetkovich, *Real Heat*, 17–39; Carole Pateman, *The Sexual Contract* (Oxford: Polity Press, 1988); Caroline Ramazanoglu, *Feminism and the Contradictions of Oppression* (London: Routledge Press); Cynthia Cockburn, *Brothers: Male Dominance and Technological Change* (London: Pluto Press, 1983); Cockburn, *In the Way of Women*; Lynne Segal, *Slow Motion: Changing Masculinities, Changing Men* (New Brunswick: Rutgers University Press, 1990). For men's self-definitions of masculinity in America and their reaction to the feminist movement, see Gary Alan Fine, "One of the Boys, Women in Male-Dominated Settings," in Michael S. Kimmel, ed., *Changing Men* (Newbury Park: Sage, 1987), 131–147). See also Sylvia Walby, ed., *Gender Segregation at Work* (Milton Keynes: Open University Press, 1988).

41. Williams, *Gender Differences At Work*, 47.

42. Moccio, interview with Thomas Van Arsdale, October 1990.

43. Ibid.

44. Ibid.

45. Francine Moccio, interview with electrical contractor, member of the Electrical Contractors Association, Flushing Queens, September 30, 1990; according to my respondent electrical work is work that only men can do.

46. Moccio, interview with Veronica Rose, New York City.

47. Moccio, interview with Harry Kelber, former Director at the Joint Industry Board, 1998.

48. Moccio, interview with Evan Ruderman, June 1, 1989.

49. Moccio, interview with Harry Van Arsdale Jr., September 18, 1985; Rorabaugh, *The Craft Apprentice*; Clawson, *Constructing Brotherhood*.

50. Moccio, interview with former president of the IBEW, Local 3, October 16, 1989.

51. Francine Moccio, interview with Maura O'Grady, former electrical apprentice, World Trade Center, New York City, June 13, 1989.

52. Francine Moccio, interview with Melinda Hernandez, June 19, 1989. For relationships between the genders in blue-collar families, see the classic account of communication between men and women, *Worlds of Pain: Life in the Working Class Family* by Lillian Rubin (New York: Basic Books, 1976). In it Rubin draws on years of research, writing, and counseling about marriage and the family, and on interviews with more than two hundred couples. Rubin explains not just how the differences between women and men arise but how they affect intimacy, sexuality, dependency, work, and so on. In "Intimate Strangers" Rubin tries to shed light on some of the reasons that men yearn for an intimate relationship which always eludes them.

53. Moccio, interview with Evan Ruderman, June 1, 1989.

54. Francine Moccio, interview with Harriet Shoup electrician apprentice, Hunter College, April 5, 1989.

55. Francine Moccio, interview with Susan Wexler, former intermediate electrician journeywoman, West Bank Café, New York City, June 2, 1989.

56. Moccio, interview with Harry Van Arsdale Jr; Moccio, interview with Thomas Van Arsdale, October 1990.

57. Francine Moccio, interview with Elvira Macri, former electrician apprentice, World Trade Center, New York City, June 20, 1989.

58. Moccio, interview with Harriet Shoup, April 5, 1989.

59. Moccio, interview with Evan Ruderman, June 1, 1989.

60. Male attitudes in the industry are hardly unique to electricians. Many features of male electricians' culture bear a striking resemblance to the cultural elements of other male settings, including elementary and middle school boys' groups (see Thorne, 1993; Everhart, 1983) and other male-dominated work and social settings (see Jean R. Schroedel, *Alone in a Crowd: Women in the Trades Tell Their Stories* [Philadelphia: Temple University Press, 1985]; Molly Martin, *Hard-Hatted Women: Stories of Struggle and Success in the Trades* [Seattle: Seal Press, 1988]; Martin, 1988; Mary Walshok, *Blue-Collar Women: Pioneers on the Male Frontie* [New York: Anchor Press, 1981]; Cynthia Cockburn, *Brothers: Male Dominance and Technological Change* [London: Pluto Press, 1983]; Carol Chetkovich, *Real Heat, Gender and Race In Urban Fire Service* [New Brunswick, NJ: Rutgers University Press, 1997]); see also Irene Padavic and Barbara Reskin, *Women and Men At Work* (Philadelphia: Temple University Press, 2002) for an excellent analysis of blue-collar work production as a "metaphor for maleness."

61. Moccio, interview with Evan Ruderman, June 1, 1989.

62. Christine Williams demonstrates this for the U.S. Marine Cops in her book, *Gender Differences at Work*. She discusses the functionality of the cadences that emphasize manliness in opposition to womanliness. Men who are weak in the Corps are also "woman-baited" to humiliate and discipline them.

63. Moccio, interview with former president of Local 3, October 16, 1989; Moccio, anonymous interviews with electrical contractors, October 10, 1991; September 30, 1990; Moccio, interview with with former official of the JIB, April 2, 1992.

64. Moccio, interview with Harry Kelber, 1998.

65. See *Annette Streeter and Ivette Ellis v. IBEW* NY, 767 F. Supp. 520 (U.S. Dist. 1991) wherein the court found that the interests of the union and the joint industry board were collusive.

66. Moccio, interview with Harry Kelber, 1998.

67. Francine Moccio, anonymous interview with electrical contractor, September 12, 1990.

68. Francine Moccio, interview with Harry Kelber, 1998.

69. Francine Moccio, interview with Colby Zieglar-Bonds, former electrician apprentice/traveler, World Trade Center, New York City, June 27, 1989.

70. Moccio, interview with Michael Wood, 1999.

71. Wider Opportunities for Women testimony from Cynthia Marano to The Commission of the Future of Worker-Management Relations, July 25, 1994.

72. Moccio, interview with Evan Ruderman, June 1, 1989.

73. Marc Bendick, Denise Hulett, Sheila Thomas, and Francine Moccio, "Enhancing Women's Inclusion Into Firefighting,"

74. Chetkovich, *Real Heat*; Ellen Rosell, Kathy Miller and Karen Barber, "Firefighting Women and Sexual Harassment," 1995, 344.

75. For case studies on gender and language, see Bingham and Gensler, *Class Action*. This book was made into a full-length feature film starring Cherlize Theron. Bingham and Gansler observe the increased use of sexually explicit language among male mineworkers when women workers are present; Chetkovich (1996, 186) also observed that white male firefighters, while less likely to legitimize racial harassment, commonly defend the use of sexually explicit language and material . Writers such as Sadker and Sadker (1994) discuss the greater acceptance of harassing behavior toward girls by boys and argue that if these were applied to racial or ethnic differences they would be unacceptable. In addition, Goodenough (1987, 440) argues that sexual harassment has a devastating effect on girls, and that if we would apply this behavior to children of another race, we would predict the outcome as negative. For an anthropological text book on the relationship of gender and language in anthropology, see Jennifer Coates, *Women, Men, and Language* (New York: Longman, 2006); Deborah Tannen, *Talking from 9 to 5—Women and Men in the Workplace: Language, Sex, and Power* (New York: Harper/Collins, 1995).

76. Moccio, interview with Maura O'Grady, June 13, 1989.

77. Francine Moccio, interview with Denise Bulger, former electrician apprentice, United Tradeswomen Conference, New York City, April 4, 1989.

78. Williams (*Gender Differences*, 1989) explains that an effort is made to maintain the gender differences in the marines in order to discourage homosexuality by segregating women and emphasizing their femininity (e.g., women marines are required to wear makeup and female uniforms, and are instructed in makeup, poise, and etiquette).

79. Moccio, interview with Evan Ruderman, June 1, 1989; Lillian Rubin discusses this issue as well in her classic work, *Intimate Strangers*.

80. Francine Moccio, interview with Evan Ruderman, June 1, 1989. It is interesting to trace the history of men's emotional fears regarding feelings for one another. Institutions from Lord Baden Powell's Boy Scouts to the denigration of women in army training all point to the use of woman-baiting as a form of institutional social control. Psychoanalytic theories of gender posit that masculinity is achieved through the rejection and denial of femininity. As psychiatrist Robert Stoller puts it, "The first order of business in being a man is: don't be a woman" (Stoller 1985: 183).

81. Cynthia F. Epstein, *Deceptive Distinctions: Sex, Gender and the Social Order* (New Haven: Yale University Press, 1988), 94. Epstein gives an example of the construction of an occupational identity among cable splicers that is a source of solidarity for some but excludes others, as its solidarity relations come at the direct expense of other workers—especially women—who are discouraged from entering the fold. The splicers, like the electricians, embrace the ethos of manliness well beyond what might reasonably be viewed as typical in manual work. This manliness is a mantra that serves as a strategic action to defend their craft against potential interlopers. Jones (1986: 293) suggests that the problem of "cultural racism" compounds the structural disad-

vantages of black people in white–dominated society. See also Henley and Kramarae (1991) on the notion of "cultural dominance" as opposed to "cultural difference." Antonio Gramsci gives the most thorough analysis of how the dominant economic and political majority in social settings and society at large influence the formation of what Gramsci terms as "cultural hegemony," that is, cultural values and social institutions which evolve into conventional wisdom even beyond the interests of its own class group.

82. Williams, *Gender Differences*.

83. Segal, *Slow Motion* (updated Palgrave 2006).

84. Women Electricians to President, Duke Electrical Supply Co., Inc. New York, NY, February 12, 1987; Women Electricians to Thomas Van Arsdale, New York, NY, February 12, 1987.

85. Ibid.

86. Catherine Mackinnon, "Pornography is Oppression," *The Ethical Spectacle* (1995), available at www.spectacle.org/1195/mack.html.

87. Moccio, interview with Beth Schulman, August 15, 1992.

88. Francine Moccio, interview with Shannon Spence, journeywoman, Empire State College, September 30, 1989.

89. Ibid.

90. Ibid.

91. Wasserstron (1977, 589, 590) notes that the ideology of sex, as opposed to the ideology of race, is a good deal more complex and confusing and as a result does not unambiguously proclaim the lesser value attached to being female rather than being male, nor does it unambiguously correspond to the existing social realities. For these reasons, among others, sexism could plausibly be regarded as a deeper phenomenon than racism. It is more deeply embedded in the culture and thus less visible.

92. Francine Moccio, interview with Shirley Merriman-Patton, Flushing Queens, July 17, 1989.

93. Ibid.

94. Moccio, interview with Colby Zieglar-Bonds, June 27, 1989.

95. Agreement and Working Rules between New York Electrical Contractors Association, Inc. and Local Union No. 3, IBEW, AFL-CIO, June 13, 2004.

96. Francine Moccio, interview with Earline Fisher, carpenter member of the District Council of Carpenters, Cornell Conference Center, New York City, October 15, 2005.

97. Ibid.

98. Ibid.

99. Ibid.

100. At the international level, the "caucus" of tradeswomen electricians are primarily women of color who have joined forces with the minority men's issues in the brotherhood.

101. Moccio, interview with Harry Kelber, 1998.

102. Ibid.

103. Ibid.

104. U.S. Bureau of the Census, Managerial and Professional Specialty Occupations, 1980. Washington, DC; U.S. Bureau of the Census, *Population and Labor Force Status by Race, Hispanic Origin, Gender for New York State 2000*, Washington, DC.

105. Francine Moccio, interview with former assistant to the chairman of the Joint Industry Board of the Electrical Industry, The Joint Industry Board, Flushing Queens, April 2, 1992.

106. Project labor agreements that are pre-hire agreements with labor organizations to utilize unionized labor for building work are currently required per a Presidential Executive Order on federal construction projects of $25 million or over.

CHAPTER 5. RACE FOR THE BROTHERHOOD

1. In Clawson, *Constructing Brotherhood*.

2. MacLean, *Freedom Is Not Enough*, 261–286, 295–299.

3. Milkman, *Gender at Work*, 158.

4. William Wilson's "wage competition" theory is that minority workers pose a threat to white male workers due to their lower labor costs. The skills mismatch hypotheses emphasizes two issues: (1) the supposedly fast changing requirements of jobs from the industrial to the postindustrial city, and (2) the failure of educational attainment among blacks to meet the demand for white collar jobs. Specifically, the decline in manufacturing in New York, for instance, decimated the "traditional goods-processing industries that once constituted the economic backbone of cities, and provided entry-level employment for lesser-skilled African-Americans."

5. Milkman, *Gender at Work*, 157.

6. Reskin and Roos, *Job Queues, Gender Queues*, 10–21. A queuing perspective suggests that labor markets consist of labor queues and job queues. Labor queues consist of all possible workers in a "queue" to fill a particular job, and the employer determines the order of the male and female workers in this queue. This determination is based on a few factors: 1) job growth and the demand for more workers in addition to males; 2) sex-specific factors such as fulfillment of antidiscrimination legislation, e.g., Title VII; 3) deterioration of industry job-related conditions leading to an undersupply of male labor and an increase in female labor.

7. Reskin and Roos, *Job Queues, Gender Queues*, 305.

8. Waldinger, *Still the Promised City?* 25, 101, 176.

9. Ray Marshall, University of Texas, Negro Participation in Apprenticeship Programs (June 1, 1963); see also U.S. Labor Department testimony from Herbert Hill on blacks in construction, October 21, 1970, quoted in MacLean, *Freedom Is Not Enough*, 91.

10. MacLean, *Freedom Is Not Enough*, 42; Whitney M. Young Jr., *To Be Equal In New York* (New York: 1964).

11. Waldinger, *Still the Promised City*, 176–179. For the full text of the Fitzgerald Act which legislated oversight of apprenticeship by federal agencies, see: www.vec .virginia.gov/docs/generaldocs/documents/fitzgerald_act_(national_apprenticeship_ program).doc, July 1, 1937.

12. Ibid., 189.

13. Ibid., 194.

14. *Apprenticeship on the Move: A System in Review: The Story of the Electrical Industry's Training Program*, Management-Labor Board publication, 1976.

15. Most notably Local 3 with the Workers' Defense League in New York City; and Local 22 in Seattle, Washington, see Ray F. Marshall and Vernon Briggs, Jr., *The Negro and Apprenticeship* (Baltimore: Johns Hopkins Press, 1967).

16. Grace Palladino, *Skilled Hands, Strong Spirits A Century of Building Trades History*, 163–164.

17. Testimony of Herbert Hill, National Labor Director, National Association for the Advancement of Colored People (NAACP) before assistant secretary of labor for wage and hour standards, Arthur A. Fletcher, at a hearing on blacks in the unionized sector of the construction industry titled, "Racial Practices of Organized Labor," October 21, 1970, Washington, D.C.

18. Thomas Bailey and Roger Waldinger, *Labor Force Adjustments in a Growing Construction Market: The Metropolitan Area in the 1990s*, Report Prepared for the Planning and Development Department of Port Authority (1987).

19. U.S. Bureau of the Census, *Census of Population and Housing, 2000*, Washington, DC.

20. Testimony of Amy Petersen, Director of Nontraditional Employment for Women, New York City Council Hearing on the Status of Women in the Building Trades, September 14, 2007, New York City.

21. New York State Department of Labor, "Labor Force by Federal EEO Occupational Group," www.labor.state.ny.us/workforceindustrydata/eeo_pr2.asp?reg=nys& geog=01000036New%20York%20State (accessed June 2007); New York State Census, Table 217, "Detailed Occupation of the Experienced Civilian Labor Force and Employed Persons by Sex" New York. 34-401-34-411.

22. For instance, according to Tri-State Census Data, in the 1970s, the small number of women in the construction trades were primarily white, and the share of black women was a mere 8.3%: that is, out of the 2,400 women that were in the skilled building trades, only 200 were black (see Appendix C, Table 3).

23. Waldinger, *Still the Promised City?* See also Joe R. Feagin and Nikitah Imni, "Racial Barriers to African American Entrepreneurship: An Exploratory Study," *Social Problems* 41, no. 4 (1994) at http://links.jstor.org/sici?sici=0377791%28199411%2941 %3A4%3C562%3ARBTAAE%3E2.0CO%3B2-g;

24. New York State Department of Labor, "Labor Force by Federal EEO Occupational Group"; Table 217, "Detailed Occupation."

25. Table 217, "Detailed Occupation."

26. Thomas Bailey and Roger Waldinger, "Primary, Secondary, and Enclave Labor Markets," *American Sociological Review* 56 (August 1991); Bailey and Waldinger, *Labor Force Adjustments*; Thomas Bailey and Roger Waldinger, *The Increasing Significance of Race In the Construction Industry,* Report Prepared for the Port Authority (1985); Waldinger, *Still the Promised City?*

27. Waldinger, *Still the Promised City?*186, 194; NOW Legal Defense and Education Fund compliance review reports courtesy of Helen Neuborne, former Executive Director. NOW retrieved the information under the Freedom of Information Act from the Department of Labor, NYS Apprenticeship Council.

28. Moccio, interview with Harry Van Arsdale Jr., September 18, 1985, on the shortening of the work day to five hours, and contractors' subsequent demand for apprentice over journeymen labor; for an excellent journalistic account of the eight-day walkout of 9,000 Local 3 construction electricians in January 1962, which resulted in Van Arsdale's winning a twenty-five-hour work week from the Electrical Contractors Association (the shortest work week ever negotiated by an American union), see A. H. Raskin, "The Big Boss of the Short Day," *New York Post,* May 12, 1962; also see *New York Sunday Post,* August 1965; and A.H. Raskin, "No 25-Hour Week for Him," *New York Times Magazine,* March 18, 1962, Section 6. The *New York Times,* usually an admirer of Van Arsdale, criticized Local 3's strike for the twenty-five-hour work week for its "defiance of all the laws of economic logic."

29. MacLean, *Freedom Is Not Enough,* 290.

30. Ibid., 219, 290.

31. George Meany, who was the former president of Local 2 of the Plumbers in New York City, also held the post of president of the AFL-CIO. Local 2 has an infamous history as one of the most notorious racially exclusionary locals in New York City. Meany once told White House domestic advisor John Ehrlichman, "When I was a plumber, it never occurred to me to have niggers in the union."

32. Rick Perlstein, "The Myths of McGovern," *Democracy: A Journal of Ideas,* Winter 2008.

33. Waldinger, *Still the Promised City?* 187; Peter Brennan, the conservative labor leader who engineered the 1970 hard hat riots in downtown Manhattan against protesting students, had been appointed by Nixon as president of the New York City Building and Construction Trades Council, and later served as Secretary of Labor under Presidents Nixon and Ford.

34. These unions removed their race bar in the mid to late 1960s, see the National Industrial Conference Board, *Handbook of Union Government, Structure, and Procedures* (1955, 63–64) in Benjamin W. Wolkinson, *Blacks, Unions, and the EEOC: A Study of Administrative Futility* (Massachusetts: Lexington Books, 1973), 146.

35. See The City of New York Commission on Human Rights, *Bias in the Building Industry, 1963–1967,* 1967, 23. See also State Advisory Committees to the U.S. Commission on Civil Rights, *Report on Apprenticeship,* 1964, 117; see also Wolkinson, *Blacks, Unions, and the EEOC,* 147.

36. Moccio, interview with Harry Van Arsdale, September 18, 1985, Charles Evans Hughes High School, New York City; Raskin, "No 25-Hour Week for Him."

37. Moccio 7, 23; see also State Advisory Committees to the U.S. Commission on Civil Rights, *Report on Apprenticeship,* 1964, 117 in Wolkinson, *Blacks, Unions, and the EEOC,* 147.

38. *Fortune Magazine* article, "The Industry Capitalism Forgot" (1947) quoted in Waldinger, *Still the Promised City?* 188. The *Fortune Magazine* article is also quoted in MacLean, *Freedom Is Not Enough,* 93–99.

39. Testimony of Herbert Hill.

40. Raskin, "No 25-Hour Week for Him."

41. Ray Marshall, University of Texas, Negro Participation in Apprenticeship Programs, June 1, 1963.

42. Ibid.

43. Raskin, "No 25-Hour Week for Him."

44. Ray Marshall, University of Texas, Negro Participation in Apprenticeship Programs, June 1, 1963.

45. Ibid.

46. "AFL and CIO Back Negro Labor Group," "Building Trade Unions Adopt Four-Point Program to Fight Discrimination," "Equal Opportunity Held 'Basic Remedy' for Birmingham," "Famed Inventor's Grandson Attends Latimer Dance," *Electrical Union World,* February 1, 1962; "For Jobs and Freedom," "Members Tell of Ten-Day Visit to Virgin Islands and Jamaica," "Outtley Appointed To Committee On Exploitation," *Electrical Union World,* May 1, 1965; "Van Arsdale, Other Union Leaders Pay Final Tribute To Reverend King At Funeral Services In Atlanta, Ga," *Electrical Union World,* April 15, 1968; "Van Arsdale Praises Rev. King For Leadership In Civil Rights Movement," *Electrical Union World,* April 1, 1964.

47. "Latimer Group Makes First Scholarship Grant," *Electrical Union World,* October 1, 1961; "Van Arsdale Receives Urban League Award," *Electrical Union World,* December 1, 1963.

48. Waldinger, *Still the Promised City?*

49. EEO-2 Data on Local 3, "Total Apprentices Accepted in Programs of Local 3 (1962–1972): A Breakdown According to Race and Gender," Local 3, "Non-White Participation in Apprenticeship Program in Selected Building Trades Unions Since March 1963," obtained through the FOIA, New York State Department of Labor, New York City Apprenticeship Council Contract Compliance Reports. EEOC Reports Joint Apprentice Committee Statistics, Local 3, "Total Apprentices Accepted in programs of Local 3 (1974–1988): A Breakdown According to Race and Gender," EEOC Reports Joint Apprentice Committee Statistics, Local 3, "Number of Apprentices Placed by the RTP," see also, "Average Annual Construction Employment in Port Authority Region (1970–1984)," New York State Department of Labor; "Annual Construction Employment New York/New Jersey Port Authority Region (1970–1984)," New York State Department of Labor. EEOC Reports Joint Apprentice Committee Statistics, Local 3, "New York and New Jersey Region Construction Contracts 1972–1985," New York State Department of Labor.

50. EEOC Reports Joint Apprentice Committee Statistics, Local 3, "Average Annual Construction Employment in Port Authority Region (1970–1984)," New York State Department of Labor Compliance Reports.

51. Julian Bond, "Hills to Climb and Strength to Climb Them," speech, Local 3 Apprentice Graduation, New York, January 15, 1973.

52. "Unionists To Set Up Scout Exhibition," *Electrical Union World*. On scouting and Local events, see also Moccio, interview with Edward J. Cleary, May 27, 1993.

53. MacLean, *Freedom Is Not Enough*, 41. See also Waldinger, p. 188, who notes the links of the Workers' Defense League to the Socialist Party in New York City. The A. Philip Randolph Institute (APRI) was cofounded in 1965 by A. Philip Randolph, President of the Sleeping Car Porters union, labor activist, and the father of the modern civil rights movement; and Bayard Rustin, civil rights leader and organizer of the 1963 March on Washington for Freedom and Jobs. The institute, which is headed today by an African American woman, Clayola Brown, has had a historic role in advancing anti-discrimination legislature and was instrumental in organizing demonstrations to place black men on construction jobs in minority communities in New York City.

54. Marshall and Briggs Jr., *The Negro and Apprenticeship*.

55. Comprehensive Employment and Training Act (CETA) of 1973, Pub. L. No. 93-203, 87 Stat. 839 (1973) [hereinafter CETA]. In 1982, CETA was replaced by the Job Training Partnership Act, Pub. L. No. 97-300, 96 Stat. 1322 (1982) (codified as amended at 29 U.S.C. & 42 U.S.C. 602 (1988). Both CETA and its successor statute were designed to finance programs that prepare youths, unskilled adults, economically disadvantaged persons, and other people facing serious barriers to employment for entry into the paid labor force. CETA also required that special consideration be given to people especially disadvantaged because of the length of time that they had been unemployed and their poor prospects for finding employment (CETA 205 © 7). Francine Moccio, interview with James Haughton, transcript, December 2007.

56. RTP prided itself on the fact that by August 1972 the organization had placed 418 black male apprentices with building trades unions, between 1972 and 1973 that number had decreased to 93; EEO-2 Forms, Equal Employment Opportunity Commission, Washington, DC, obtained at the Wagner Labor Archives, New York University; Moccio, interview with Ed Cleary, the former President of the NYS AFL-CIO discussed the aim of Local 3 to bring minority apprentices into the industry through the mentorship of community organizations. The founder of Fightback believes that RTP was merely a public relations front for the building trades, Moccio, interview with the former director of RTP,.

57. Moccio, interview with James Haughton, December 21, 2006; Waldinger, *Still the Promised City?*; Joshua B. Freeman, "The Fiscal Crisis," in *Working Class New York* (New York: The New Press, 2000).

58. *Santiago Iglesias Club Educational Society: A Beginning*, online article (accessed July 24, 2007).

59. Moccio, interview with Stan Brown, journeyman electrician, August 1990.

60. Ibid.

61. Ibid.

62. Moccio, interview with Lafayette "Buddy" Jackson, March 15, 1990 and April 2, 1992.

63. Moccio, interview with James Haughton, December 21, 2006.

64. Eventually, according to an African American journeyman who wishes to remain anonymous, "when Harlem Fightback—which was a hybrid of Old and New Left politics—split into a variety of advocacy groups in New York, high level board members of those groups were placed on overseeing jobs, like Kennedy Airport, they are paid off and then the contractors can say they have black representation." These splintered

organizations came under the control of white mafia who calmed the racial fears of contractors by extorting money from them to keep the labor peace on construction sites. The system they used was based on the inner-city grassroots community coalitions run by influential individuals and infiltrated by the Gambino crime family. New York City's legendary coalitions are white mafia male Teamsters who used blacks to instigate incidents on job sites, then offered services to the contractor to keep blacks out. White leaders of the coalition got jobs, and some blacks were rewarded by being placed on a job site. But blacks have had to pay cash to white Teamster extortionists every week out of their paycheck (see Selwyn Raab, "Extortion Cases Expose Mob Ties to Minority-Hiring Groups," *New York Times*, November 10, 1999). The article reports that activities of coalitions on construction sites are carried out under the guise of advocacy organizations aiming to integrate minority men into unionized building trades. But individuals and coalition mafia extort kickbacks from both contractors and workers they have placed on job sites.

65. Moccio, interview with James Haughton, December 21, 2006.

66. New York City Commission on Human Rights on Construction Industry Discrimination, 1989; U.T.B. United Third Bridge, Inc., v. IBEW NY, 512 F. Supp. 288 (U.S. Dist. 1981).

67. Waldinger, *Still the Promised City?* 188.

68. Moccio, interview with James Haughton, December 21, 2006.

69. Waldinger, *Still the Promised City?* 205.

70. Ibid., 214–215. Waldinger discusses the struggle of Jewish tradesmen to integrate into electrical work and other building trades in the 1920s in New York City. See also "Electrical Welfare Club Marks 61 Years of Progress," *Electrical Union World*, November 15, 1984; *Electrical Welfare Club, Inc., 50th Golden Anniversary*, publication of the JIB, April 7, 1973; Moccio, interview with Bernard Rosenberg, March 20, 1990.

71. Moccio, interview with Stan Brown, August 1990.

72. Moccio, interview with Lafayette "Buddy" Jackson, March 15, 1990 and April 2, 192l; Moccio, interview with James Haughton, December 21, 2006.

73. Waldinger, *Still the Promised City?* 291–293.

74. Moccio, interview with James Haughton, December 21, 2006.

75. MacLean, *Freedom Is Not Enough*, 42.

76. Moccio, interview with James Haughton, , December 21, 2006.

77. Waldinger, *Still the Promised City?* 183, 189.

78. Ibid.

79. Moccio, interview with James Haughton, December 21, 2006.

80. Ibid.

81. MacLean, *Freedom Is Not Enough,* 92, 94.

82. Moccio, interview with Lafayette "Buddy" Jackson, March 15, 1990 and April 2, 1992.

83. See "Van Arsdale Praises Rev. King For Leadership in Civil Rights Movement," *Electrical Union World*, April 1, 1964; "Van Arsdale Receives Urban League Award," *Electrical Union World*, December 1, 1963; "Local 3 Gives $1,000 to Aid Reverend King," *Electrical Union World*, December 1, 1960.

84. Waldinger, *Still the Promised City?* 189, 195.

85. MacLean, *Freedom Is Not Enough*, 97.

86. Ibid., 91–95.

87. Ibid., 95–97.

88. Waldinger, *Still the Promised City?* 192–195.

89. Moccio, interview with Edward J. Cleary, May 27, 1993.

90. Moccio, interview with Lafayette "Buddy" Jackson, March 15, 1990, and April 2, 1992.

91. New York City Commission on Human Rights, *Discrimination in New York City Construction Trades* (December 1993).

92. *Apprenticeship on the Move*, 1976.

93. *A Portrait of Minority Participation in the Electrical Industry*, booklet published by the Joint Industry Board of the Electrical Industry, Harry Van Arsdale Jr. Library, Empire State College (SUNY), 1993.

94. New York City, Office of Construction Industry Relations, Problems of Discrimination and Extortion in the Building Trades (New York: Office of Construction Industry Relations, Mayor's Office, 1982), 6; see also Waldinger, *Still the Promised City?* p. 357; Emmanuel Perlmutter, "Building Unions Defend Hiring Policy," *New York Times*, March 10, 1971; *New York Times*, July 20, 1973.

95. Moccio, interview with Edward Cleary, May 27, 1993.

96. Francine Moccio, interview with Mary Au, electrician journeywoman, Charles Evans High School, June 1, 1989.

97. Schuck v. SDHR et. al., 102 AD2d 673, 1984.

98. Moccio, interview with James Haughton, December 21, 2006.

99. Waldinger, *Still the Promised City?*

100. For the most prominent lawsuit brought by the EEOC on behalf of black and Hispanic members of Local 28 for Local 28's noncompliance with court-ordered decisions to remedy past discrimination in recruitment and admission into the union, see *Equal Employment Opportunity Commission v. Local 638 ETC*. The case was finally settled in 2004, after thirty years of litigation. Other lawsuits against a broad variety of building trades, such as ironworkers, steamfitters, plumbers, and electricians, for race and sex discrimination can be found at The Office of Construction Industry Relations, Problems of Discrimination in New York City's Construction Trades. See also Waldinger, *Still the Promised City?* 195.

101. Waldinger, *Still the Promised City?* 192, 194. See also Palladino, *Skilled Hands*, 143, 170.

102. *Matter of Monarch Electrical Corporation v. Roberts*, 70 NY 2d 91, 517 NYS 2d 711; also see Waldinger, *Still the Promised City?* 191–196. See also MacLean, *Freedom Is Not Enough*, 95.

103. Cook, "Local 300," 113–143.

104. The pattern of ethnic layering and its relationship to the development of ethnic niches in manufacturing and construction work in New York City by twentieth-century immigrant groups are discussed at length in Waldinger, *Still the Promised City?* 94–136.

105. Ashley Carlins, "Electrical Square Club Stresses Good Citizenship," *Electrical Union World*; Moccio, interview with Edward J. Cleary, May 27, 1993.

106. *Santiago Igleasias Club Educational Society: A Beginning*, accessed July 24, 2007; "Farmed Inventor's Grandson Attends Latimer Dance," *Electrical Union World*, February 1, 1962; "Latimer Group Makes First Scholarship Grant," *Electrical Union World*, October 1, 1961.

107. "Chinese American Cultural Society Hosts Labor Festival," *Electrical Union World*, November 24, 1999.

108. Moccio, interview with former president of the Lewis Howard Latimar Society, Harlem, New York City, May 15, 2006.

109. Moccio, interview with Harry Van Arsdale Jr., September 18, 1985

110. Moccio, interview with Stan Brown, August 1990.

111. Moccio, anonymous interview with black male journeyman, HVA School for Labor Studies, Electrical Apprentice Program, Charles Evans Hughes High School, Empire State College (SUNY), September 15, 1991.

112. U.S. Bureau of the Census, *Managerial and Professional Specialty Occupations, 1980*, Washington DC; U.S. Bureau of the Census, *Managerial and Professional*

Specialty Occupations, 1990, Washington DC; U.S. Bureau of the Census, *Occupation Detailed Code List, 2000*, Washington DC; U.S. Bureau of the Census, *Population and Labor Force Status by Race, Hispanic Origin, Gender for New York State 2000*, Washington, DC.

113. Moccio, interview with former president of the Lewis Howard Latimar Society, Harlem, New York City, May 15, 2006.

114. *Apprenticeship on the Move.*

115. Moccio, interview with James Haughton, December 21, 2006; Moccio, interview with Lafayette "Buddy" Jackson, March 15, 1990 and April 2, 1992.

116. Moccio, interview with James Haughton, December 21, 2006.

117. New York City Commission on Human Rights, *Discrimination in New York City Construction Trades* (December 1993).

118. Compliance reports on apprenticeship and EEO guidelines obtained courtesy of NOW Legal Defense and Education Fund.

119. Agreement and Working Rules between NY Electrical Contractors Assoc., Inc., Assoc. of Electrical Contractors, Inc., and Local Union No. 3, IBEW, June 10, 1983.

120. Moccio, Field Notes, April 8, 1992.

121. Moccio, interview with former president of Local 3, October 16, 1989; Moccio, interview with Howard Hirsch, October 10, 1991; Moccio, interview with former official of the JIB, April 2, 1992.

122. Bailey and Waldinger, "Primary, Secondary, and Enclave Labor Markets."

123. Ibid.; EEO-2 Data on Joint Industry Board of the Electrical Industry, Washington, DC.

124. Waldinger, *Still the Promised City?*; MacLean, *Freedom Is Not Enough*. For the dire consequences of the fiscal crisis of the 1970s in New York to tradesmen, see Miriam Pavel, "On a 'Breadline' for Skilled Jobs," *Newsday*, 1975.

125. Waldinger, *Still the Promised City?*.

126. Ibid.

127. There is a 'myth' in the Local among field electricians that one of the most powerful families in the Local were "actually breeding male electricians" in order to step up their control over the Local.

128. Christopher Eriksen is the current business manager and the grandson of Harry Van Arsdale as well as a member of the McBurn Family in the brotherhood.

129. Moccio, interview with Edward J. Cleary, May 27, 1993. See also, Moccio, interview with Bernard Rosenberg, March 20, 1990. For a history of ethnic layering in the building trades regarding northern European immigrants such as the Scots, Irish, and Germans, see Waldinger, *Still the Promised City?* 22, 23, 41, 42, 201, 204–205.

130. Cook, "Local 300," emphasized the function of social clubs in Local 3 as "political precincts."

131. Moccio, interview with Harry Van Arsdale Jr., September 18, 1985.

132. Father to son sponsorship began to wane in the post-World War II period as European-Americans began to fill more professional occupational niches in New York City. The minority clubs were established to address this problem of recruiting prospective young men into the apprenticeship program who could be sponsored by a member in good standing of the brotherhood and disciplined informally in case of behavioral problems on work sites.

133. Moccio, Interview field notes.

134. Young black males were formerly recruited through outreach programs like RTP and the WDL. The demise of these training programs for minority males had a negative effect on the supply of young black men available for apprenticeships. In addition, this also meant fewer black male journeymen in the Local. This impacts the number of blacks as small business electrical contractors (as journeymen invariably become

small contractors at some point in their careers). According to black journeymen interviewed for this study, blacks feel that there is substantial discrimination against minority-owned businesses in general, and especially in construction, as well as other federal and state funded building and infrastructural projects, such as bridge, tunnel, and road work in New York City. For a comparison with New York on the progress of minority-owned business on federal and state subsidized work, see John Lunn and Huery L. Perry, "Justifying Affirmative Action: Highway Construction in Louisiana," *Industrial and Labor Relations Review*, 46, no. 3 (April, 1993), 464–479. For an analysis of the length of apprenticeship and its effect on integrating black male workers, see George Strauss, "Minority Membership in Apprenticeship Programs in the Construction Trades: Comment," *Industrial and Labor Relations Review*, 27, no. 1 (Oct., 1973), 93–99.

135. Johnny Dwyer, "This Hammer for Hire," *New York Times*, February 4, 2007.

136. Waldinger, *Still the Promised City?*

137. Ibid., 274.

138. Joshua B. Freeman, "Hardhats: Construction Workers, Manliness, and the 1970 Pro-War Demonstrations," *Journal of Social History* 25, no. 4 ( ); H. Ron Davidson, Harry Van Arsdale and the New York City Fiscal Crisis (unpublished paper), August 1999.

139. Waldinger, *Still the Promised City?*

140. Moccio, anonymous interview with electrical contractor, October 10, 1991; Moccio, anonymous interview with former director of apprenticeship program, March 15, 1990 and April 2, 1992.

141. Moccio, interview with Edward J. Cleary, May 27, 1993; Moccio, interview with James Haughton, December 21, 2006; Moccio, interview with former official of the JIB, April 2, 1992.

142. Moccio, anonymous interview with black male electrician, member of the Lewis Howard Latimar Society, Harlem, New York City, 2007.

143. Moccio, interview with Bernard Rosenberg, March 20, 1990.

144. Francine Moccio, interview with Mary Ellen Boyd, transcript, March 1992.

145. Mary Corcoran, Irish Illegals, p. 57, quoted in Waldinger, *Still the Promised City?* 204.

146. For example, in Local 3, the Allied club, the Catholic Council, as well as the Westchester Mechanics previously discussed.

147. Waldinger, *Still the Promised City?*

148. James Haughton believes that the Workers' Defense League was merely a public relations front to show governmental action but it was not effective at recruiting black men into apprenticeship. In the 1980s, minority males accepted in the electrical construction apprenticeship program as new registrants dropped precipitously from 27.7 percent to 15.4 percent; even though, paradoxically, Census data from the same period show a slight increase of approximately 2 percent of minority male labor participation rates (Table 5, except for apprentices). Waldinger, *Still the Promised City?* 188–190, also makes the claim that the WDL never really had the interests of black workers at its core.

149. Thomas Bailey and Roger Waldinger, paper for Port Authority, cite; New York State compliance reviews obtained from the New York State Department of Labor for the apprenticeship program via FOIA indicate that the decline of the WDL workers' training programs negatively impacted the recruitment of black and minority men into the Local's apprenticeship.

150. Moccio, interview with Edward J. Cleary, May 27, 1993; Moccio, interview with Lafayette "Buddy" Jackson, March 15, 1990 and April 2, 1992.

151. U.T.B United Third Bridge, Inc., v. IBEW New York, 512 F. Supp. 288. U.S. Dist. 1981.

152. Samuel Lopez, an electrician of Puerto Rican descent, and Charles Calloway, a black male electrician, both members of Local 3, filed the lawsuit. There were also members of the United Bridge, which was a militant minority male organization that demonstrated on construction sites in order to get contractors to hire men of color. There is no family relations between Samuel Lopez and Jose Lopez, one of the first Puerto Rican electricians in New York City, who headed the Santiago Iglesias club in the late 1950s, and his son Edwin Lopez, his son and a Local 3 electrician who is also an active leader in Iglesias.

153. The outcome is significant for affirmative action. This is a class action suit (www.jjccsettlement.com/) brought on behalf of black and Latin carpenters, freight handlers, and others.

154. NYC Commission on Human Rights, Questions for the Commissioners, 1989.

155. NYC Commission on Human Rights, April 24, 1990.

156. Ibid., 37.

157. Based on an examination of Census data from 1970 to 2000, as well as apprenticeship rates for New York State for 2007 obtained from the New York State Department of Labor.

158. MacLean, *Freedom Is Not Enough*, 261–286, 295–299.

159. Kessler-Harris, *In Pursuit of Equity*, 280; for a discussion of the especially complex situation of black women regarding benefits of affirmative action policies, see Kessler-Harris, 270.

160. "Civilian Employed Population 16 Years and Over, Data Set," American Community Survey 2006. From the late 1970s through the decade of the 1990s and into the twenty-first century, journalists have reported extensively on women's struggle to integrate the New York City building trades. See, for example, Sherry Dean, "Constructing a Strong Future," *Daily News*, October 15, 1981; Claudia H. Deutsch, "Getting Women Down to the Site," *New York Times*, March 11, 1990.

161. Moccio, interview with Mary Ellen Boyd, March 1992.

162. Moccio, interview with former director of apprenticeship program, March 15, 1990 and April 2, 1992.

163. EEOC Data on Electrical Construction Workers in the Tri-State Area of New York, New Jersey, and Connecticut from 1970 to 2000; additional tables will be made available; for 2005 Labor Force Participation Rates for Women and Men, go to www.dol.gov/wb/factsheets/Qf-ESWM05.htm

164. See Table 2 in Appendix B.

165. Francine Moccio, anonymous interview with male electrician, March 29, 2003.

166. Moccio, interview with former director of apprenticeship program, March 15, 1990 and April 2, 1992; Moccio, interview with former president of Local 3, October 16, 1989. See also Susan Eisenberg, "Women Hard Hats Speak Out," *The Nation*, September 18, 1989. Also, for the slow pace of women's integration into building trades' jobs during the decade of the 1990s, see Lisa Genasci, "Women Find Work in Traditionally Male Jobs But Inroads Being Made At Snail's Pace As Employees Meet Masculine Resistance" *Chicago Tribune*, March 20, 1995.

167. New York City Human Rights Commission, Hearings Transcripts, 1963, 1990, and 1993.

168. Moccio, anonymous interview with electrical contractor, October 10, 1991.

169. Moccio, interview with Earline Fisher, October 15, 2005.

170. Annette Streeter and Ivette Ellis v. IBEW NY, 767 F. Supp. 520. U.S. Dist. 1991.

171. Ibid.

172. Moccio, interview with Harry Kelber, 1996.

173. Ibid.

174. Moccio, interview with former president of Local 3, October 16, 1989; Moccio, interview with former business manager of Local 3, October 1990.

175. Moccio, interview with former business manager of Local 3, October 1990.

176. Moccio, interview with Michael Wood, male journeyman electrician.

177. Francine Moccio, anonymous interview with male journeyman electrician, March 29, 2003.

178. Moccio, interview with Evan Ruderman, June 1, 1989.

179. Nancy Chodorow, "Family Structure and Feminine Personality," quoted in Clawson, *Constructing Brotherhood*.

180. Moccio, interview with Earline Fisher, October 15, 2005.

181. Kessler-Harris, *In Pursuit of Equality*, for a detailed discussion of assertions of black civil rights leaders and their testimony before national legislators on how the restoration of black masculinity compelled them to give priority to issues of race over gender in applying civil rights laws to challenge exclusionary patterns of hiring.

182. MacLean, *Freedom Is Not Enough*, for an in-depth discussion of twentieth-century workplace practices to exclude black men from skilled occupations based on white prejudicial notions of their inability to acquire job skills.

183. Moccio, interview with anonymous black journeywoman electrician.

184. Ibid.

185. Moccio, interview with Stan Brown, August 1990.

186. It appears that Hispanic men have been the most successful in integrating into the electricians' trade and brotherhood. This may also be due to the rapid increase in the number of Hispanics into the New York City labor market as a whole. Nonetheless, the 'unofficial story' among tradesmen and tradeswomen attributes their success to Hispanics' tight informal networks of communication about job opportunities as well as contracting work in the industry. Hispanics are reported to have developed a high degree of informal kinship networks sponsoring family members into the trade. See also Waldinger, *Still the Promised City?* 298. This does not imply that Hispanics do not experience racism in seeking jobs. An analysis comparing patterns of hiring outcomes of young Anglo and Hispanic job-seekers shows that in local firms that lack affirmative action policies, young Hispanic males are less likely to be hired than their Anglo counterparts. See Genevieve M. Kenney and Douglas A. Wissoker, "An Analysis of the Correlates of Discrimination Facing Young Hispanic Job-Seekers," *The American Economic Review*, 884, no. 3 (June, 1994), 674–683. Hispanics also face discrimination in the New York City building trades (see the Human Rights Commission Hearings on Discrimination in the New York City Building Trades, 1993). Black men have made much slower gains than their Hispanic counterparts in the trade. Waldinger attributes this to black men pursuing public employment jobs rather than seeking construction apprenticeships post-Civil Rights era. In addition, he further argues that black men (as well as leaders of the Civil Rights Movement) were not very successful (mainly due to racism) in developing small contracting firms once they achieved their journeyman cards. Asian male electricians interviewed for this book feel they have made some gains in the industry and trade but mostly in the field, not as electrical contractors. Similar to their black male counterparts, Asian men face the challenge of severe racism and exclusion in the larger society, which also mitigates against access to the informal networks in construction that are necessary to gaining information and bidding opportunities for building contracts.

187. *Apprenticeship on the Move.*

188. At the time (1993) of my first survey of women electrician pioneers in Local 3, there were 295 women in Division A. Fifteen years later, in 2008, tradewomen electrician activists report that the Local has 300 women electricians (apprentices and journeywomen) in the Division A.

CHAPTER 6. A CLUB OF HER OWN

1. Francine Moccio, interview with Evan Ruderman, June 1, 1989.
2. Ibid.
3. Ibid.
4. Ibid.
5. Moccio, interview with Cynthia Long, May 5, 1992, and October 30, 2003.
6. Mary Wollstonecraft quoted in Clawson, *Construction Brotherhood*.
7. *A Portrait of Minority Participation in the Electrical Industry*, booklet, publication of the Joint Industry Board of the Electrical Industry, Harry Van Arsdale School for Labor Studies, Empire State College (SUNY), 1993.
8. Martha Midgett, *Our Industry's Mosaic*, publication of the Joint Industry Board of the Electrical Industry, 1978. Midgett was the first president of the fraternal club, the Women's Active Association of Local 3 representing women of color in the electrical manufacturing divisions. For a detailed discussion of the role of reform labor feminists to the male leadership of unions, see Cobble, *The Other Women's Movement*, 14–19.
9. Martha Midgett, *Our Industry's Mosaic*, publication of the Joint Industry Board of the Electrical Industry, 1978. Midgett was the first president of the fraternal club, the Women's Active Association of Local 3 representing women of color in the electrical manufacturing divisions.
10. Midgett, *Our Industry's Mosaic*. See also Cobble, *The Other Women's Movement*. For an authoritative account of the role of working class women in union leadership, see Kessler-Harris, *In Pursuit of Equity*.
11. Ibid.
12. "Local 3 Honors Scroll Ladies," "Member Tell of Ten-Day Visit to Virgin Islands and Jamaica," and "N.Y. City Labor Leaders Urge Shop Stewards Help Build Brotherhood Party," *Electrical Union World*, October 15, 1961.
13. "30 More Mfg. Division Members Finish Blue Print Reading Class," *Electrical Union World*, June 1971.
14. A "punchlist" is a contract document of "to do" items used in architecture and the U.S. building trades.
15. "Shape-up" is a term first coined by dockworkers who were day laborers seeking work on the New York City piers. The term has been adapted to classify per diem workers across a broad range of industries and occupations.
16. Moccio, interview with Cynthia Long, May 5, 1992; Moccio, interview with Melinda Hernandez, June 19, 1989; Moccio, interview with Evan Ruderman, June 1, 1989; Moccio, interview with Beth Schulman, 1993; Moccio, interview with Jackie Simpson, electrician journeywoman, October 1991.
17. Moccio, interview with Evan Ruderman, June 1, 1989.
18. Moccio, interview with Melinda Hernandez, June 19, 1989.
19. Moccio, interview with Cynthia Long, May 5, 1992.
20. Moccio, interview with Laura Kelber, February 20, 1990.
21. The conference was conducted by Cornell University faculty at the Institute for Women and Work.
22. Moccio, interview with Evan Ruderman, June 1, 1989.
23. Ibid.
24. Ibid. See also Susan Faludi, *Stiffed: The Betrayal of Modern Man* (Chattow & Windus, 1999) and *Backlash: The Undeclared War Against American Women* (New York: Anchor Book, 1992). In *Stiffed*, Faludi sees men as victims of declining economic changes which have eliminated male-typed skilled manufacturing trades' jobs.
25. For the development of male identity and occupational culture, see also Chetkovich, *Real Heat*, 155, 124–130, on firefighters, and Bingham and Gansler, *Class Action*, 121, on male identity and the occupation of mineworkers.

26. Moccio, interview with former president of Local 3, October 16, 1989.

27. Moccio, interview with Evan Ruderman, June 1, 1989; Moccio, interview with Cynthia Long, May 5, 1992; Moccio, interview with Melinda Hernandez, June 19, 1989.

28. Moccio, interview with anonymous officer of Local 3.

29. Moccio, with Harry Kelber, 1996.

30. Moccio, interview with Evan Ruderman, June 1, 1989. See also Kris Paap (*Working Construction*, 2006) for a discussion of male carpenters and union loyalty which offers a contrast with male electricians in Local 3. The former view association with the carpenters' union as a loss in status, whereas male electricians embrace opportunities to affiliate with the electrical brotherhood and its clubs and activities. Male electricians, especially white men with family members in Local 3 are, for the most part, strong unionists who cling to the mythologies associated with the brotherhood's founding and view it as monolithic. Men of color who have been able to develop a line of familial succession in the Local share similar views, regarding the union "as a life raft."

31. Moccio, interview with Cynthia Long,

32. Moccio, interview with Evan Ruderman, June 1, 1989.

33. Moccio, interview with Stan Brown, August 1990.

34. Moccio, interview with Cynthia Long, May 5, 1992.

35. Moccio, interview with Melinda Hernandez, June 19, 1989.

36. Moccio, interview with Cynthia Long, May 5, 1992.

37. Moccio, interview with Laura Kelber, February 20, 1990.

38. Francine Moccio, interview with Betty Friedan, Washington, DC, June 14, 2005.

39. Moccio, interview with Laura Kelber, February 20, 1990.

40. Ibid.

41. Francine Moccio, interview with Cheryl Farrell, former president of the Amber Light Society, Cornell Conference Center, New York City, July 16, 2005.

42. Moccio, interview with Stan Brown, August 1990.

43. Moccio, interview with Melinda Hernandez, June 19, 1989.

44. Janet L. Rogers, *Lesbian Women in the Building Trades: Conference Work for American Women's Labor History*. Unpublished PhD diss. (May 1990).

45. Moccio, interview with Emily Forman, June 26, 1989.

46. Moccio, interview with Shirley Merriman-Patton, July 17, 1989.

47. Moccio, interview with Evan Ruderman, June 1, 1989.

48. Moccio, interview with Cynthia Long, May 5, 1992.

49. Moccio, interview with Cynthia Long, May 5, 1992, Cornell University, 2003.

50. Moccio, interview with Shirley Merriman-Patton, July 17, 1989.

51. Francine Moccio, interview with Dorothy Mays, electrician journeywoman, Flushing Queens, June 8, 1989.

52. Francine Moccio, telephone interview with Donna Simms, Division A electrician, MIJ, May 22, 1989.

53. Black women respondents are not only more economically disadvantaged than white women (see Malveaux and Wallace, 1987), but also report a more complex set of experiences on work sites. They feel extremely vulnerable as possible targets of violence by men—both white and black but for different reasons. Case studies in the literature on race and gender oppression at work confirm the double disadvantage women of color, especially black women in nontraditional jobs such as firefighters (Chetkovich, 1999), 185.

54. Moccio, interview with Shirley Merriman-Patton, July 17, 1989. Although some black women felt more connected to the Local in a minority fraternal club, they were unable to voice their problems regarding issues of sex discrimination on the job site in these meetings.

55. Francine Moccio, telephone interview with Donna Simms electrician MIJ, May 22, 1989.

56. Moccio, interview with Mary Au, June 1, 1989.

57. Moccio, interview with Melinda Hernandez, June 19, 1989.

58. Moccio, interview with Evan Ruderman, June 1, 1989.

59. Governmental enforcement for compliance regulations are very weak; for example, even though the carpenter's union in New York City has been put into trusteeship by the Justice Department because of corruption and racketeering charges, governmental agencies have never challenged their failure to comply with affirmative action goals. Enforcement in the construction industry, as a whole, has been very weak since Clarence Thomas took over the EEOC in 1981, and shifted the oversight for affirmative action to the joint industry and labor-management boards, rather than the EEOC agency itself.

60. Moccio, interview with Evan Ruderman, June 1, 1989.

61. Moccio, interview with former president of Local 3, October 16, 1989.

62. Moccio, interview with Edward J. Cleary, May 27, 1993.

63. Moccio, anonymous interview with union leader, New York City.

64. Moccio, interview with Cynthia Long, May 5, 1992; Moccio, interview with Evan Ruderman, June 1, 1989.

65. Moccio, interview with Elvira Macri, June 20, 1989.

66. Ibid.

67. Moccio, Field Notes, Bayberry Land Weekend Conference for Union Leaders of Local 3 and Female Electricians, February 29, 1996.

68. Moccio, interview with Veronica Rose,; Moccio, interview with Cynthia Long, May 5, 1992; Moccio interview with Melinda Hernandez, June 19, 1989; Moccio, interview with Niki Basque, electrician journeywoman and first female shop steward in Local 3's construction division, 1999.

69. The Local promoted a Latina and mother of five children as the first foreperson in the union and active spokeswoman for the Local. Reportedly, she had previously threatened to sue the Local for discrimination and an assault against her from a male coworker at a New York City construction site.

70. Moccio, interview with Melinda Hernandez, June 19, 1989.

71. Moccio, interview with Melinda Hernandez, June 19, 1989.

72. Women Electricians to President, Duke Electrical Supply Co., Inc., New York, February 12, 1987.

73. Laura Kelber to Thomas Van Arsdale, Brooklyn, NY, January 14, 1989; Letter from Beth Goldman, Cynthia Long, Ruth Strasman, Laura Kelber, and Alice Tyson-Carver, New York; Letter to Local Negotiating Committee, February, 1989; Women Electricians to Thomas Van Arsdale, New York, February 12, 1987.

74. Moccio, interview with Cynthia Long, May 5, 1992.

75. Ibid.

76. Moccio, interview with Evan Ruderman, June 1, 1989.

77. Ibid.

78. Ibid.

79. Moccio, interview with Laura Kelber, February 20, 1990.

80. Moccio, interview with Cynthia Long, May 5, 1992; Moccio, interview with Melinda Hernandez, June 19, 1989; Francine Moccio, interview with Evan Ruderman, June 1, 1989; Moccio interview with Beth Schulman, 1993.

81. Letter from Women Electricians (WE) to Local's Negotiating Committee, February 1989.

82. Moccio, interview with Cynthia Long, May 5, 1992.

83. Moccio, interview with Evan Ruderman, June 1, 1989.

84. Ibid.

85. Moccio, interview with Melinda Hernandez, June 19, 1989.

86. Moccio, interview with Susan Wexler, June 2, 1989.

87. Moccio, interview with Tony Pecorello, January 11, 1991.

88. Moccio, interview with Mary Ellen Boyd, transcript, March 1992; New York City Commission on Human Rights on Construction Industry Discrimination, 1989; Wider Opportunities for Women Testimony from Cynthia Marano to The Commission of the Future of Worker-Management Relations, July 25, 1994.

89. New York City Commission on Human Rights on Construction Industry Discrimination (hearings) (New York City: One Police Plaza, 1990); Moccio, interview with James Haughton, December 21, 2006.

90. Francine Moccio, interview with Jane Latour, former director of the Women's Rights Project at the Association for Union Democracy (AUD), February 1, 2001.. AUD has been successfully spearheading an attempt to rid the industry of corruption and racketeering, which stand in the way of women's progress as well as men's.

91. New York City Commission on Human Rights on Construction Industry Discrimination, 1989.

92. Tom Robbins, "Rackets Remedy," *Village Voice*, December 14, 2005.

93. Moccio, interview with Evan Ruderman, June 1, 1989.

94. Leaders of the Women's Club, women electricians were prominent in the founding of Operation Punchlist.

95. Focus Group Sessions of women electricians, May 31, 1990; March 15, 1998; May 22, 2003; June 30, 2006; May 3, 2008.

96. Moccio, interview with Cheryl Farrell, July 16, 2005.

97. Ibid.

98. Francine Moccio, anonymous interview with female electrician journeywoman, member of the Amber Light Society, Cornell Conference Center, New York City, July 16, 2005.

99. Francine Moccio, anonymous interview with white male journeyman electrician, 2005.

100. Moccio, interview with Cheryl Farrell, July 16, 2005.

101. Anonymous Focus Group Sessions, women electricians, May 31, 1990; March 15, 1998; May 22, 2003; June 30, 2006; May 3, 2007.

102. Francine Moccio, interview with anonymous journeywoman, New York City, May 10 and 11, 2007.

103. Francine Moccio, anonymous interview with female journeywomen, currently president of The Amber Light Society, 2006.

104. Ibid. An amber (fossilized resin) light indicates that electrical power has been supplied to the equipment. A green light means that a signal is present. A red light indicates a fault or error.

105. Francine Moccio, anonymous interview with female electrician, October 12, 2007.

106. Ibid.

107. Francine Moccio, anonymous interview with female electrician, October 17, 2007.

108. Ibid.

109. Ibid.

CONCLUSION

1. Leon F. Bouvier and Vernon M. Briggs Jr., *The Population and Labor Force of New York: 1990–2050* (Washington, DC, Population Reference Bureau, 1988); see also American Community Survey and Data 2006.

2. Chetkovich, *Real Heat*; and Clara Bingham and Laura Leedy Gansler, *Class Action: The Story of Lois Jenson and the Landmark Case that Changed Sexual Harassment Law* (New York: Doubleday, 2002).

3. See Dan Bell, "Women Not Wanted: Female Construction Workers Face Chronic Unemployment and Daunting Odds: A New Mayoral Commission Will Have to Change the Face of an Industry," *City Limits* (May/June 2005), 12–15. For nontraditional jobs, see Jerry Jacobs, *Revolving Doors: Job Sex Segregation and Women's Careers* (Stanford: Stanford University Press, 1990), as well as for a discussion of the gender wage gap and the integration of men and women, and blacks and whites at work.

4. Judy Mann, "Job Discrimination Still A Menace," *The Washington Post*, Dec 12, 1997, E.03. See also Lisa Genasci, "Women Find Work in Traditionally Male Jobs But Inroads Being Made at Snail's Pace as Employees Meet Masculine Resistance," *Chicago Tribune*, March 20, 1995, 8. See also Ileen A. DeVault, "We'll Call You If We Need You: Experiences of Women Working Construction," by Susan Eisenberg, [Review], *The Journal of American History*, 86, no. 1 (June 1999), 331–332.

5. See Gary Becker, *The Economic Approach to Human Behavior* (1976), quoted in Jacobs, *Revolving Doors*. Although discriminatory intent and disparate treatment in employment discrimination cases are difficult to prove, the number of sex and race discrimination complaints have grown (as reported by the EEOC). There also have been recent legal cases naming both employers and unions as defendants in sex discrimination cases, see the following: Priscilla Villines, Plaintiff, v. United Brotherhood of Carpenters and Joiners of America, AFL-CIO, Defendant, Civil Action No. 9601886 (RMU), Document Nos.: 21, 25, 28 & 31; Rosa Elizabeth Hunt, Plaintiff, v. Robert Weatherbee, ET AL., Defendants, Civil Action No. 84-3001-Y, U.S. District Court For The District of Massachusetts, 626 F. Supp. 109 1986 U.S. Dist. LEXIS 30131; 121 L.R.R.M. 2408; 39 Fair Empl. Prac. Cas. (BNA) 1469; 39 Empl. Rac. Dec. (CCH) P35,927; Danuta Ryduchowski, Plaintiff,-against-The Port Authority Of New York And New Jersey, Defendant. 96-CV-5589 (JG), U.S. District Court For The Eastern District of New York, 1998.

6. See Courtney Derwinski, "'Math At Work' Women in Nontraditional Careers: Women in Dentistry, and Women in Engineering," *Labor Studies Journal* 30, no. 3 (Fall 2005), 87–89. See also Michael B. Katz, "Women and the Paradox of Economic Inequality in the Twentieth Century," *Journal of Social History* 39, no. 1 (Fall 2005), 65–88; Jenrose Fitzgerald, "Gendering Work Reworking Gender: The Contested Terrain of 'Women and Work,'" *Journal of Women's History* 12, no. 3 (Fall 2000), 207–217.

7. Joseph Pleck, *The Myth of Masculinity* (Cambridge: MIT Press, 1981). Pleck argues that masculine values are more vigorously reinforced in the socialization process of men as compared with feminine values for women; thus steering men into "manly" jobs.

8. Mastracci, *Breaking Out of the Pink-Collar Ghetto*. Mastracci argues that career counseling and employment and training policies greatly influence the number of women seeking blue-collar jobs in skilled occupations.

9. David Gordon, Richard C. Edwards, and Michael Reich, *Segmented Work, Divided Workers: The Historical Transformation of Workers in the United States* (New York: Cambridge University Press, 1982). See also Jeffrey Waddoups and Djeto Assane, "Mobility and Gender in a Segmented Labor Market: A Closer Look," *The American Journal of Economics and Sociology* (October, 1993).

10. Milkman, *Gender at Work*, 68–69.

11. Ibid., 102–103.

12. Quoted in Milkman, *Gender at Work*, 102.

13. The District Council of Carpenters has recently launched a public relations campaign to appeal to all male members, and especially minority male and female members, to save the residential building market for unionized labor by fighting the growth of substandard wages and union labor in the residential construction market in New York City. See "Let's Fight to Take Back the Residential Market in New York!"

The Construction Workers' News Service, 2008. For a critique of this effort, see also Gregory A. Butler, "New York District Council of Carpenters Sells Out Non Union Residential Carpenters," http://finance.groups.yahoo.com/group/gangbox/

14. See Bingham and Gansler, *Class Action The Landmark Case That Changed Sexual Harassment Law* (New York: Random House, 2004) for an account of women in mining; Reskin and Roos (*Job Queues, Gender Queues,* 1990) for women in the computer industry; Milkman (*Gender at Work,* 1989) on auto manufacturing; Venus Green, *Race On the Line: Gender, Labor, and Technology in the Bell System 1890–1980* (Durham, NC: Duke University Press, 2001) on telecommunications.

15. For an excellent assessment of women's experiences in federally and state mandated building trades apprenticeships, see Lynn Judith Shaw, *Women Union Electricians: A Comparison of Job and Training Experiences of White Women and Women of Color*, unpublished masters theses, California State University, Long Beach, Ca., 1995.

16. Waldinger, *Still the Promised City?* 205.

17. The electrical brotherhood and the electricians' trade is influenced by the Catholic doctrine of a tradesman's right to a "living wage' developed by American Catholic thinkers and defined as "reasonable hours and fair conditions," quoted in Montgomery (*Fall of the House of Labor*, 1987), 308. In addition, Catholic thinkers in the first half of the twentieth century professed the doctrine that if women who are wives and mothers engaged in paid work, the welfare of the whole family and society would suffer. Christian men believed that the wife who becomes a wage worker is no longer a wife. (Ryan, "Living Wage," in Montgomery, *Fall of the House of Labor*, 308n133.

18. As stated previously, in the first half of the twentieth century, women electrical workers in factories were depicted as having "dexterity" and "nimble fingers," see Schatz, *The Electrical Workers*, 30.

19. The Local and the Joint Industry Board currently have fifty-four supply houses in and around the New York metropolitan area.

20. Here, I am defining paternalism that promotes the notion that women, like children, are inferior creatures whom men must take care of. See references to the work of Mary R. Jackman, *The Velvet Glove: Paternalism and Conflict in Gender, Class, and Race Relations* (Berkeley: University of California Press, 1994) and L. H. Sanders, "Efficiency and Women Employees," *Mass Transportation: City Transit's Industry-Wide Magazine*, 39(7) (July 1943): 244, 257 on the definition of paternalism as "protection" of women and children, and attitudes of male employers as quoted in Padavic and Reskin (*Women and Men at Work*, 2002, 42). For a detailed explanation of the way in which notions of "protection" for women and children factor into the historical development of capitalism and patriarchy, see Gerder Lerner, *The Creation of Patriarchy (Women and History)* (New York: Oxford University Press, 1986).

21. Clawson, *Constructing Brotherhood*, 8.

22. Hobsbawm, *Workers*, 66–82, 259–259.

23. For an excellent description of social clubs in the needle trades, see Will Herberg, "The Old-Timers and the Newcomers: Ethnic Relations in a Needle Trades Union," *Journal of Social Issues* 9, no. 1 (1953): 32–39; on the topic of working women's social clubs in the late nineteenth and early twentieth century, see Priscilla Murolo, *The Commonground of Womanhood, Class, Gender, and Working Girls' Clubs, 1884–1928* (Urbana and Chicago: University of Illinois Press, 1997); also see the "traveling" tradition of cigar makers in Patricia A. Cooper, *Once A Cigar Maker*, 75–89.

24. David Horowitz, *Ethnic Groups in Conflict* (Berkeley: University of California Press, 1985).

25. MacLean, *Freedom Is Not Enough*, 94.

26. Jacobs, *Revolving Doors*, 178.

27. MacLean, *Freedom Is Not Enough*, 42.

28. Ibid., 43–44

29. Rorabaugh (*The Craft Apprentice*, 1986, 9) discusses the role of the black slave in the craft trades prior to and after the Civil War both on the Southern plantations and in Northern metropolitan centers like New York. With regard to the influx of new immigrants, the IBEW just recently passed a resolution to criminalize undocumented immigrant labor in the United States and to restrict immigration. This position stands in contradiction to the resolution of the AFL-CIO passed at their national convention which recommends amnesty for undocumented immigrant workers; however, the IBEW opposes amnesty for illegal immigrants. See also, "Shipyard Workers Organize to Stop 21st Century Slavery," *Workday Minnesota*, March 11, 2008, www.workdayminnesota.org/index.phy?news-6-3550. Indian workers from Bombay working for Signal International in the aftermath of Hurricane Katrina in New Orleans walked off the job in March 2008, throwing their hardhats into the air, and singing, "We Shall Overcome." Placards which read, "I Am a Man," were used in picketing the company that has recruited workers from India only and, according to workers and the New Orleans Workers' Center for Racial Justice, treat them as slaves.

30. See Wilentz, *Chants Democratic*, 264; Rorabaugh, *The Craft Apprentice*, 181; Montgomery, *Fall of the House of Labor*, 24–25. For an analysis of white privilege and racism in the wages of tradesmen, see the excellent work of David R. Roediger, *Wages of Whiteness: Race and the Making of the American Working Class* (New York: Verso Books 1991).

31. Montgomery, *The Fall of the House of Labor*, 87–111.

32. Ibid.

33. See John Schmitt and Ben Zipperer, "The decline of African-Americans in union and manufacturing, 1979-2006," in *The Black Commentator*, available at www.blackcommentator.corm/220/220-cover-decline-african-americans-unions-mfg.html. The share of Hispanics rose from 5.8 percent in 1983 to 11.5 percent of all union workers in 2006.

34. New York State Apprenticeship Council Statistics, Report to the Subcommittee on Minority Advancement in the Building Trades Unions, June 1980, and June 2007. See also Waldinger, *Still the Promised City?* 137–174, for a discussion on minority participation in construction in New York City during the building boom of the mid- to late-1980s.

35. "Voluntary agreements" such as the initial agreement Van Arsdale made to induct minority male apprentices have been referred to as "more publicity than progress" by Herbert Hill, the former NAACP Labor Secretary, and "no substitute for enforcement," quoted in MacLean, *Freedom Is Not Enough*, 94.

36. Generally, Hispanics have made greater inroads as well into unionized jobs nationally.

37. Respondents commented that "Joe-six pack;" that is, typical white male construction workers, are infamously anti-immigrant, and their unions have adopted resolutions criminalizing illegal immigrant workers.

38. Chetkovich, *Real Heat*. See also Marc Bendick, Denise Hulett, Sheila Thomas, and Francine Moccio, "Enhancing Women's Integration in Firefighting in the U.S.A," *The International Journal of Diversity in Organizations, Communities, and Nations* (July 2008). For harassment in private sector jobs, see Bingham and Gansler, *Class Action*. For an analysis of the cultural resistance women police officers encounter by their male counterparts in private sector versus public sector work, see Kathryn Scarborough, *Women in Public and Private Law Enforcement* (Boston: Butterworth-Heinemann, 2002). For a comparative perspective on women in policing in the U.S. and Great Britain, see Jennifer M. Brown, *Gender and Policing: Comparative Perspectives* (Houndmills: Macmillan, 2000). For glass ceiling issues women confront in policing, see Dorothy Moses Schulz, *Breaking the Brass Ceiling: Women Policy Chiefs and their Paths to*

the Top (Westport, CT: Praeger Publishers, 2004). For the important role women today play in a variety of nontraditional jobs, see Susan Hagan, *Women at Ground Zero: Stories of Courage and Compassion* (Indianapolis: Alpha Books, 2002).

39. The Victoria King incident is quoted in Lynn Judith Shaw, Women Union Electricians: A Comparison of Job and Training Experiences of White Women and Women of Color, unpublished PhD diss. (California State University, Long Beach, 2002); Shaw cites reference to Victoria King's book, *Manhandled Black Females* (Nashville, TN: Winston-Derek Publishers, 1992).

40. Interview with Donna Simms, 1995 quoted in Lynn Judith Shaw, *Women Union Electricians: A Comparison of Job and Training Experiences of White Women and Women of Color*, unpublished Ph.D. diss., California State University, Long Beach, 2002.

41. Unlike male contractors interviewed for this book, women electrical contractors I interviewed, who were also former Local 3 journeywomen, encouraged more women into the trade and ensured equal opportunity to women when hiring for their subcontracted electrical work projects on building sites.

42. Ibid., 102.

43. One male contractor interviewed for this book told me that it's "to be expected."

44. Francine Moccio anonymous interview with male electrician journeyman, Case #1M, Cornell University, New York City, May 15, 2003. For a discussion of the effects of harassment on women in nontraditional work, see Barbara Gutek, *Sex and the Workplace: The Impact of Sexual Behavior and Harassment on Women, Men, and Organization* (San Francisco: Jossey-Bass, 1985). Gutek observes that men's work environments tend to be much more sexualized than are women's. See Catherine MacKinnon, "Pornography is Oppression," and Susan Brownmiller, , work on the use of sexual harassment and pornography as a way to pressure women to leave their jobs. See also Peggy E. Bruggman, "Beyond Pinups: Workplace Restrictions on the Private Consumption of Pornography," *Hastings Constitutional Law Quarterly* 23 (Fall 1995), 271–311. For a discussion of the use of sexual language as harassment, see Gary Alan Fine, "One of the Boys: Women in Male-Dominated Settings," in *Changing Men: New Directions in Research on Men and Masculinity*, ed. M.S. Kimmel (Newbury Park, CA: Sage Publications, 1987).

45. According to the majority of tradeswomen interviewed for this book, men across a broad spectrum of trades have anonymously communicated their disgust at women's treatment on the job sites. Nonetheless, men lack the moral courage to speak up against this abuse by foremen, coworkers, and managers fearful that they will not be called back to the job. For similar reactions of men to women in nontraditional occupations, see also Chetkovich, *Real Heat*, 77–82; in police work, see Alexandra Gillen, *Equality and Difference in the Evolution of Women's Police Role*, PhD diss. (University of Chicago, 2003), 336 pages, AAT 3077057. For oral history accounts of men's reactions to women in construction, see Eisenberg, *We'll Call You When We Need You*, and Martin, *Hard-hatted Women: Stories of Struggle and Success in the Trades* (Seattle: Seal Press, 1988). Men's behavior toward women in the broader context of nontraditional jobs is the subject of numerous books and articles, for a sampling of this work, see Bridget O'Farrell and Sharon L. Harlan (1982) and Padavic and Reskin, *Job Queues, Gender Queues, Explaining Women's Inroads into Male Occupations* (Philadelphia, PA: Temple University Press, 1990). Padavic and Reskin analyzed men's behavior and women's interest in nontraditional blue-collar work, and the intersection of race and gender in shaping both women's and men's responses to workplace harassment and abuse (*Women and Men at Work*, 68, 70). See also Stephanie M. Wildman, *Privilege Revealed: How Invisible Preference Undermines America* (New York: New York University Press, 1996), 90–99 (Appendix A in this book has additional useful references on this topic pertaining to women in nontraditional blue-collar jobs).

46. Christine L. Williams, *Gender Differences at Work: Women and Men in Non-traditional Occupations* (Berkeley: University of California Press, 1989). Traditionally, many male apprentices are former veterans of the military. The New York Building Trades Council has a present-day program to recruit young men just out of military service into the electrical trade called, "From Helmets to Hard Hats," see Palladino, *Skilled Hands*, 229, 230. But the lack of outreach to female veterans on the part of male leaders is disappointing.

47. For a discussion of the tradition and ethical code of workers' "manly bearing" in construction to "endure hardship" and assist a brother on the job site; see Montgomery, *The Fall of the House of Labor*, 295–297.

48. Eliot Brown, "Help Wanted: The Shortage of Labor In Construction Threatens Safety," *The New York Observer*, June 2008. For the special situation of immigrant workers in New York's construction industry, see the U.S. Department of Labor, Occupational Safety and Health Administration, February 2002, online www.osha.gov/pls/oshaweb/owadisp.show_document?p_table=TESTIMONIES&p_id=286

49. Bingham and Gansler, *Class Action*, 88.

50. Moccio, interview with Melinda Hernandez, October 29, 1999.

51. Moccio, interview with Evan Ruderman, December 5, 2001.

52. Moccio, interview with Cynthia Long, June 25, 2002.

53. Ibid.

54. Moccio, anonymous interview with Local 3 female electrician journeywoman, Case #015, April 10, 2006.

55. Stephanie M. Wildman, *Privilege Revealed: How Invisible Preference Undermines America* (New York: New York University Press, 1996), 30–31.

56. Clawson, "Introduction," in *Constructing Brotherhood*.

57. Ibid.

58. Christopher Lasch, *Haven in a Heartless Work: The Family Besieged* (New York: Basic Books, 1995).

59. MacLean, *Freedom Is Not Enough*, 299. MacLean acknowledges the relationship of gender relations at home to the treatment of women in the construction workplace.

60. They are also fearful for their jobs which have sustain them with a middle-class lifestyle in spite of the fact that they lack higher education. This is especially true of white men who are unionized in the trade. They are acutely aware of their privileged position vis-à-vis their white male counterparts in other blue-collar occupations and industries. And they are determined to keep it that way. Fearful that immigrants, minorities, and especially women, as well as other white males may replace them, men in the industry cling to the Brotherhood like a life raft in a surging sea of union shops and economic uncertainty.

61. Francine Moccio, anonymous interview with male electrician journeyman, Electrical apprenticeship program in New York City, October 15, 1998.

62. For a discussion of the ways in which fraternities on university campuses inhibit the equality of male and female students, see Mindy Stombler, "'Buddies' or 'Slutties': The Collective Sexual Reputation of Fraternity Little Sisters," *Gender and Society*, 8, no. 3 (Sept., 1994), 297–323.

63. Montgomery, *Fall of the House of Labor*, 143–148. See also Priscilla Murolo, *The Common Ground of Womanhood: Class, Gender and Working Girls' Clubs, 1884–1928* (Urbana: University of Illinois Press, 1997).

64. Chetkovich, *Real Heat*, 181.

65. Moccio, interview with Evan Ruderman, June 1, 1989.

66. This also undermines the position of white workers, as divisions along lines of race and national origin undermine further the solidarity necessary to keep the union effective and growing.

67. There are a number of ongoing legal actions on behalf of women plaintiffs against large contractors in New York City, such as the Jacob Javits Center which is known for its nepotistic hiring practices and unfair treatment of workers. Many other legal battles pertaining to the rebuilding of lower Manhattan and the compliance of contractors to hire women and minorities have not even entered the courts.

68. For a historical account of the relation of women's wages to their social status, see Kessler-Harris, *A Woman's Wage*.

69. For a detailed description of the American Recovery and Reinvestment Act of 2009, see http://en.wikipedia.org/wiki/American_Recovery_and_Reinvestment_Act _of_2009.

70. For the White House Announcement of the Presidential Order for Use of Project Labor Agreements on Federal Construction Projects, see: www.whitehouse. gov/the_press_office/executiveorderuseofprojectlaboragreementsforfederalconstruc-tionprojects/.

71. *The Empire State News.Net*, "Port Authority Study Shows Significant Job Creation, Economic Activity by WTC Building," March 6, 2009.

72. Gerda Lerner, *The Creation of Patriarchy, Women and History* (New York and Oxford: Oxford University Press, 1986).

Selected References

CHAPTERS OR OTHER PARTS OF BOOKS

Chomsky, Noam. "Interpreting the World," in *The Chomsky Reader*. New York: Pantheon Books, 1987.

Cook, Alice H. "Local 300: Guided Democracy," in *Union Democracy: Practice and Ideal: An Analysis of Four Large Local Unions*. New York: Cornell University, 1963: 113–143.

Freeman, Joshua B. "The Fiscal Crisis," in *Working Class New York*. New York: The New Press, 2000.

Joyce, John T. "Codetermination, Collective Bargaining, and Worker Participation in the Construction Industry," in Thomas A. Kochan, ed. *Challenges and Choices Facing American Labor*. Cambridge, MA: MIT Press, 1986: 257.

Ladner, Joyce A. "Racism and Tradition: Black Womanhood in Historical Perspective," in Filomena C. Steady, ed., *The Black Woman Cross-Culturally*. Cambridge, MA: Schenkman, 1981: 285.

Solokoff, Natalie. "What's Happening to Women's Employment: Issues for Women's Labor Struggles in the Late 1980s–1990s," in *Hidden Aspects of Women's Work*. New York: Praeger, 1987.

BOOKS

Appier, J. *Policing Women: The Sexual Politics of Law Enforcement and the LAPD*. Philadelphia: Temple University Press, 1998.

Argersinger, Jo Ann E. *Making the Amalgamated: Gender, Ethnicity, and Class in the Baltimore Clothing Industry, 1899–1939*. Baltimore, MD: Johns Hopkins University Press, 1999.

Benson, Susan P. *Counter Cultures: Saleswomen, Customers, and Managers in American Department Stores*. Urbana: University of Illinois Press, 1986.

Briggs, Vernon M., and Felicia M. Foltman. *Apprenticeship Research: Emerging Finding and Future Trends*. Washington, DC: Library of Congress, 1981.

Briggs, Vernon M., and Ray F. Marshall. *The Negro and Apprenticeship*. Baltimore, MD: Johns Hopkins University Press, 1967.

Chetkovich, Carol. *Real Heat, Gender and Race in Urban Fire Service*. New Brunswick, NJ: Rutgers University Press, 1997.

Chodorow, Nancy. *The Reproduction of Mothering*. Berkeley: University of California Press, 1978.

Clawson, Mary Ann. *Constructing Brotherhood: Class, Gender and Fraternalism*. Princeton, NJ: Princeton University Press, 1989.

Clifford, James, and George Marcus, eds. *Writing Culture*. Berkeley: University of California Press, 1986.

Cobble, Dorothy Sue. *Dishing It Out: Waitresses and Unions in the Twentieth Century*. Urbana: University of Illinois Press, 1991.

———. *The Other Women's Movement: Workplace Justice and Social Rights in Modern America*. Princeton, NJ: Princeton University Press, 2004.

Cockburn, Cynthia. *Brothers: Male Dominance and Technological Change*. London: Pluto Press, 1983.

———. *Machinery of Dominance: Women, Men and Technical Knowhow*. London: Pluto Press, 1985.

———. *In the Way of Women: Men's Resistance to Sex Equality in Organizations*. New York: Industrial-Labor Relations Press, Cornell University, 1991.

Cook, Alice H. *Union Democracy: Practice and Ideal, An Analysis of Four Large Local Unions*. Ithaca, NY: Cornell University, 1963.

Cooper, Patricia A. *Once a Cigar Maker: Men, Women, and Work Culture in American Cigar Factories, 1900–1919*. Urbana: University of Illinois Press, 1987.

DeVault, Ileen A. *United Apart, Gender and the Rise of Craft Unionism*, Ithaca, NY: Cornell University Press, 2004.

Edwards, Richard, David M. Gordon, and Michael Reich, eds. *Segmented Work, Divided Workers: The Historical Transformation of Labor in the United States*, New York: Cambridge University Press, 1982: 288.

Epstein, Cynthia F. *Deceptive Distinctions: Sex, Gender and the Social Order*. New Haven, CT: Yale University Press, 1988.

Gilligan, Carol. *In a Different Voice: Psychological Theory and Women's Development*. Cambridge, MA: Harvard University Press, 1993.

Gramsci, Antonio. *Selections from Political Writings: 1921–1926*. International Publishers, 1978.

———. *Selections from the Prison Notebooks*. London: Lawrence and Wishart, 1971.

Hagan, Susan. *Women at Ground Zero: Stories of Courage and Compassion*. Indianapolis, IN: Alpha Books, 2002.

Harlan, Sharon L., and Ronnie J. Steinberg. *Job Training for Women: The Promise and Limits of Public Policies*. Philadelphia, PA: Temple University Press, 1989.

Hobsbawm, Eric J. *Primitive Rebels: Studies in Archaic Forms of Social Movements in the 19th and 20th Centuries*. New York: Norton Library, 1965.

Hoch, Paul. *White Hero, Black Beast*. London: Pluto Press, 1979.

Jeal, Tim. *The Boy-Man: The Life of the Lord Baden-Powell*. New York: Morrow, 1990.

Kazin, Michael. *Barons of Labor, The San Francisco Building Trades and Union Power in the Progressive Era*. Urbana: University of Illinois Press, 1989.

Kessler-Harris, Alice. *A Woman's Wage: Historical Meanings and Social Consequences*. Lexington: University Press of Kentucky, 1990.

———. *In Pursuit of Equity: Women, Men and the Quest for Economic Citizenship in 20th Century America*. Oxford: Oxford University Press, 2003.

Kimmel, Michael S., ed. *Changing Men*. Newbury Park, CA: Sage, 1987.

Kondo, Dorrine K. *Crafting Selves: Power, Gender, and Discourses of Identity in a Japanese Workplace*. Chicago, IL: University of Chicago Press, 1990.

Lerner, Gerda. *The Creation of Patriarchy (Women and History)*. New York: Oxford University Press, 1986.

MacLean, Nancy. *Freedom Is Not Enough, The Opening of the American Workplace*. Cambridge, MA: Harvard University Press, 2006.

Marshall, Ray F. *The Negro and Organized Labor*. Sydney, Australia: John Wiley, 1965.

Martin, Molly. *Hard-Hatted Women: Stories of Struggle and Success in the Trades*. Seattle, WA: Seal Press, 1988.

Mastracci, Sharon. *Breaking Out of the Pink-Collar Ghetto: Policy Solutions for Non-College Women*. Armonk, NY: M.E. Sharpe, 2004.

McWilliams, Wilson C. *The Idea of Fraternity in America*. Berkeley: University of California Press, 1973.

Milkman, Ruth. *Gender at Work: The Dynamics of Job Segregation by Sex during World War II*. Urbana: University of Illinois Press, 1988.

Montgomery, David. *Citizen Worker: The Experience of Workers in the United States with Democracy and the Free Market During the Nineteenth Century*. New York: Cambridge University Press, 1995.

———. *Fall of the House of Labor: The Workplace, the State, and American Labor Activism, 1865–1925*. New York: Cambridge University Press, 1987.

———. *Workers' Control in America: Studies in the History of Work, Technology and Labor Struggles*. New York: Cambridge University Press, 1979.

Morris, Robert. *Lights and Shadows of Freemasonry*. Louisville, KY: J.F. Brennan, 1852.

Neufeld, Maurice. *Day In and Day Out*. Ithaca, NY: Cornell University Press, 1955.

Paap, Kris. *Working Construction: Why White Working-Class Men Put Themselves in Harm's Way*. Ithaca, NY: ILR Press, 2006.

Padavic, Irene, and Barbara Reskin. *Women and Men at Work*, 2nd ed. Thousand Oaks, CA: Pine Forge Press, 2002.

Palladino, Grace. *Skilled Hands, Strong Spirits: A Century of Building Trades History*. Ithaca, NY: Cornell University Press, 2005.

Pateman, Carole. *The Sexual Contract*. Palo Alto, CA: Stanford University Press, 1988.

Pleck, Joseph. *The Myth of Masculinity*. Cambridge, MA: MIT Press, 1981.

Polanyi, Karl. *The Greatest Transformation: The Political and Economic Origins of Our Time*. Boston: Beacon Press, 1970.

Ramazanoglu, Caroline. *Feminism and the Contradictions of Oppression*. London: Routledge, 1989.

Raphael, Ray. *The Men from the Boys: Rites of Passage in Male America*. Lincoln: University of Nebraska Press, 1988.

Reskin, Barbara, and Patricia Roos. *Job Queues, Gender Queues*. Philadelphia: Temple University Press, 1990.

Rorabaugh, W. J. *The Craft Apprentice: From Franklin to the Machine Age in America*. New York: Oxford University Press, 1988.

Roseberry, William. *Anthropologies and Histories: Essays in Culture, History, and Political Economy*. New Brunswick, NJ: Rutgers University Press, 1994.

Rubin, Lillian B. *Worlds of Pain: Life in the Working Class Family*. New York: Basic Books, 1976.

Ruffini, Gene. *Harry Van Arsdale, Jr: Labor's Champion*. Armonk, NY: M.E. Sharpe, 2002.

Schatz, Ronald W. *The Electrical Workers, A History of Labor at General Electric and Westinghouse 1923–60*. Urbana and Chicago: University of Illinois Press, 1983.

Schroedel, Jean R. *Alone in a Crowd: Women in the Trades Tell Their Stories.* Philadelphia: Temple University Press, 1985.

Segal, Lynne. *Slow Motion: Changing Masculinities, Changing Men.* New Brunswick, NJ: Rutgers University Press, 1990.

Siedman, Harold. *Labor Czars: A History of Labor Racketeering.* New York: Liveright, 1938.

Snitow, Ann, Christine Stansell, and Sharon Thompson. *Powers of Desire: The Politics of Sexuality.* New York: Monthly Review Press, 1983.

Tilly, Louise A., and Joan W. Scott. *Women, Work & Family.* New York: Methuen, 1987.

Tolson, Andrew. *The Limits of Masculinity.* London: Tacistock Publications, 1977.

Walby, Sylvia. *Patriarchy at Work.* Oxford: Polity Press, 1986.

———. *Theorizing Patriarchy.* Oxford: Basil Blackwell, 1990.

———. ed. *Gender Segregation at Work.* Milton Keynes, England: Open University Press, 1988.

Waldinger, Roger. *Still the Promised City?* Cambridge, MA: Harvard University Press, 1999.

Walshok, Mary. *Blue-Collar Women: Pioneers on the Male Frontier.* New York: Anchor Press, 1981.

Wildman, Stephanie M. *Privilege Revealed, How Invisible Preference Undermines America.* New York: New York University Press, 1996.

Williams, Christine L. *Gender Differences at Work: Women and Men in Nontraditional Occupations.* Berkeley: University of California Press, 1989.

Wolkinson, Benjamin W. *Blacks, Unions and the EEOC.* Lexington, KY: Lexington Books, 1973.

CENSUS DATA

U.S. Bureau of the Census. *Census of Population and Housing, 2000,* Washington, DC.

———. *Data Dictionary: Alphabetical Index by Variable Name. (Person Record), 2000.* Washington, DC.

———. *Detailed Industry Code List, 2000,* Washington, DC.

———. *Managerial and Professional Specialty Occupations, 1980,* Washington, DC.

———. *Managerial and Professional Specialty Occupations, 1990,* Washington, DC.

———. *Occupation Classification, 1970.* Washington, DC.

———. *Occupation Detailed Code List, 2000.* Washington, DC.

———. *Population and Labor Force Status by Race, Hispanic Origin, Gender for New York State 2000.* Washington, DC.

COURT CASES

Annette Streeter and Ivette Ellis v. IBEW NY, 767 F.Supp. 520, U.S. Dist. 1991.

Darla G. Hall, Patty J. Baxter, and Jeanette Ticknow v. Gus Construction Co., Inc., 842 F.2d 1010, U.S. App. 1988.

EEOC v. Gurtz Electric Co. and Pickus Construction & Equipment Co., U.S. App. 2003.

Efrain Lebron v. IBEW Local 3, U.S. Dist. 1991.

Jonna Snapp-Foust v. National Construction, LLC, 1 F.Supp. 2d 773, U.S. Dist. 1997.

Luther E. Jones v. Intermountain Power Project, 794 F.2d 546, U.S. App. 1986.

Port Authority of NY and NJ v. Danuta Ryduchowski, 530 U.S. 1276, 120 S. Ct. 2743, 147 L. Ed. 2d 1007, U.S. 2000.

Priscilla Villines v. United Brotherhood of Carpenters and Joiners of American, AFL-CIO 999, F.Supp. 97, U.S. Dist 1998.

Thomas Waters, Jr., and John H. Rowland v. Olinkraft, Inc. and Local 2660, 475 F. Supp. 743, U.S. Dist. 1979.
U.T.B. United Third Bridge, Inc. v. IBEW NY, 512 F.Supp. 288, U.S. Dist. 1981.

DISSERTATIONS

DeVault, Ileen A. *Strong Hands to Aid the Weak*, unpublished dissertation, 1990.
Gillen, Alexandra. *Equality and Difference in the Evolution of Women's Police Role*, University of Chicago, 2003, AAT 307057.
Mastracci, Sharon. *Labor and Service Delivery: Training Programs for Women in Non-Traditional Occupations*, The University of Texas at Austin, 2001, AAT 3037525.
Rogers, Janet L. *Lesbian Women in the Building Trades: Conference Work for American Women's Labor History*, unpublished dissertation, 1990.
Santiago, George. *Union Power and Affiliation*, unpublished dissertation, 1987.
Shaw, Lynn Judith. *Women Union Electricians: A Comparison of Job and Training Experiences of White Women and Women of Color*, California State University Long Beach, 1995, AAT 1377494.

FIELD NOTES

Bayberry Land Weekend Conference for the Male Leaders and Female Electricians, February 29, 1992.
Tradeswomen Now and Tomorrow, founding meeting, Bobst Library, New York University, New York City, and the District Council of Carpenters, 2001.
United Tradeswomen 10th Anniversary Conference, New York, NY, December 3, 1989.

HEARINGS/TESTIMONIES

New York City Commission on Human Rights. *Building Barriers: Discrimination in New York City's Construction Industry*. New York: 1990 and 1993, One Police Plaza, New York.
U.S. Labor Department, testimony from Herbert Hill on Blacks in Construction, October 21, 1970.
Wider Opportunities for Women, testimony from Cynthia Marano to The Commission of the Future of Worker-Management Relations, July 25, 1994.

JOURNAL ARTICLES IN PRINT

Aniakudo, P., and J. Yoder. "Outsider Within the Firehouse: Subordination and Difference in the Social Interactions of American Women Firefighters," *Gender and Society* 11, no. 3 (1997).
Bailey, Thomas, and Roger Waldinger. "Primary, Secondary, and Enclave Labor Markets," *American Sociological Review* 56 (August 1991).
Bender, Daniel. "Too Much of Distasteful Masculinity: Historicizing Sexual Harassment in the Garment Sweatshop Factory," *Journal of Women's History* 15, no. 4 (Winter 2004), 91–116.
Byrd, Barbara, "Women in Carpentry Apprenticeship: A Case Study," *Labor Studies Journal* 24, no. 3 (Fall 1999).
Chetkovich, Carol. "Women's Agency in a Context of Oppression: Assessing Srategies for Personal Action and Public Policy," *Hypatia* 19, no. 4 (Fall 2004).
Eisenberg, Susan. "We'll Call You If We Need You: Experiences on Women Working Construction," *The Journal of American History* 86, no. 1 (1999).

———. "Women's Educational Equity Act Publishing Center Digest," *Women's Educational Equity Act Publishing Center Digest* (August 1992).

Freeman, Gordon M. "Editorial: About Apprenticeship," *Electrical Workers' Journal* (1993).

Freeman, Joshua B. "Hardhats: Construction Workers, Manliness, and the 1970 Pro-War Demonstrations," *Journal of Social History* 25, no. 4, (Summer 1993).

Genasci, Lisa. "More Women Take on Nontraditional Jobs," *Journal Record* (March 9, 1995).

Harlan, Sharon L., and Brigid O'Farrell. "Craftworkers and Clerks: The Effect of Male Co-Worker Hostility on Women's Satisfaction with Non-Traditional Blue-Collar Jobs," *Social Problems* 26, (February 1982): 252–64.

Hartmann, Heidi. "Capitalism, Patriarchy and Job Segregation by Sex," *Signs* 1, no. 3 (Spring 1976): 137–169.

Haynes, Anthony, and Helena Worthen. "Getting In: The Experience of Minority Graduates of the Building Bridges Project Pre-Apprenticeship Class," *Labor Studies Journal* 28, no. 1 (Spring 2003).

Johnson, Lynn. "Women in Construction," *Women: A Journal of Liberation* 8, no. 3 (August 1983).

Josephs, Susan L., Janina C. Latack, Bonnie L. Roach, and Mitchell D. Levine. "The Union as Help or Hindrance: Experiences of Women Apprentices in the Construction Trades," *Labor Studies Journal* (Spring 1988).

Katz, Michael B. "Women and the Paradox of Economic Inequality in the Twentieth-Century," *Journal of Social History* 39, no. 1 (Fall 2005), 65–88.

Kerr, Daniel. "Cracking the Temp Trap, Day Laborers' Grievances and Strategies for Change in Cleveland, Ohio," *Labor Studies Journal* 29, no. 4 (Winter 2005), 87–108.

Law, Sylvia A. "Voices of Experience: New Responses to Gender Discourse," *Harvard Civil Liberties Law Review* 24, no. 1 (Winter 1989).

Lee, Jo Ann, and Helen T. Palmer. "Female Workers' Acceptance in Traditionally Male-Dominated Blue-Collar Jobs," *Sex Roles* 22, no. 9/10 (1990).

Leonard, Jonathan S. "The Effect of Unions on the Employment of Blacks, Hispanics, and Women," *Industrial and Labor Relations Review* 39, no. 1 (October 1985).

Lillydahl, Jane. "Women and Traditionally Male Blue-Collar Jobs," *Women and Occupations* 13, no. 3 (August 1986).

Martin, Molly. "Affirmative Action in Big Trouble," *Tradeswomen* 1, no.2 (Summer 1981).

Mastracci, Sharon H. "Persistent Problems Demand Consistent Solutions: Evaluating Policies to Mitigate Occupational Segregation by Gender," *Review of Radical Political Economics* 37 (Winter 2005), 23.

———. "The 'Institutions' in Institutionalization, Programs for Women in Highly Skilled, High-wage Occupations," *Working USA*, April 30, 2003.

McIlwee, Judith S. "Women's Survival in Nontraditional Blue-Collar Occupations," *Frontiers* 7, no. 1 (1991).

———. "Work Satisfaction Among Women in Nontraditional Occupations," *Frontiers* 6, no. 1 (1983).

Messing, Karen. "Equality and Difference in the Workplace: Physical Job Demands, Occupational Illnesses, and Sex Differences," *NWSA Journal* 12, no. 3 (Fall 2000), 21–49.

Meyer, Steve. "Rough Manhood: The Aggressive Confrontational Shop Culture of U.S. Auto Workers During World War II," *Journal of Science History* 36, no. 1 (Fall 2002), 125–147.

Neufeld, Maurice F. "Day In, Day Out With Local 3, IBEW," *New York State School of Industrial and Labor Relations Bulletin* 17 (June 1951).

O'Farrell, Brigid. "Women in Blue Collar and Related Occupations at the End of the Millennium," *Quarterly Review of Economics and Finance*, vol. 39, issue 5, (1999) 65-88.

O'Farrell, Brigid, and Sharon L. Harlan. "Craft Workers and Clerks: The Effect of Male-Co-Worker Hostility on Women's Satisfaction with Non-Traditional Jobs," *Social Problems* 29, no. 3 (February 1982).

Padavic, Irene. "Attractiveness of Working in Blue Collar Jobs for Black and White Women: Economic Need, Exposure, and Attitudes," *Social Science Quarterly* 72, no. 1 (1991).

———. "The Re-creation of Gender in a Male Workplace," *Symbolic International* 14, no. 3 (1991).

———. "White-Collar Work Values and Women's Interest in Blue-Collar Jobs," *Gender and Society* 6, no. 2 (June 1992).

Padavic, Irene, and Barbara Reskin, "Men's Behavior and Women's Interest in Blue-Collar Jobs," *Social Problems* 37, no. 4 (November 1990).

Pearce, Diana, "The Feminization of Poverty: Women, Work, and Welfare." *Urban and Social Change Review* 11 (1978): 28–36.

Sacks, Karen. "The Class Roots of Feminism," *Monthly Review* 27 (1976).

Santos, Cecilia MacDowell. "En-gendering the Police: Women's Police Stations and Feminism in Sao Paulo," *Latin American Research Review* 39, no. 3 (2004), 29–55.

Scott, Joan, and Louise Tilly. "Women's Work and the Family in Nineteenth Century Europe," *Comparative Studies in Society and History* 17 (1975).

Suh, Doowson. "Middle-Class Formation and Class Alliance," *Social Science History* 26, no. 1. (Spring 2002), 105–137.

Swerdlow, Marian. "Men's Accommodations to Women Entering a Nontraditional Occupation: A Case of Rapid Transit Operatives," *Gender & Society* 3, no. 3 (September 1989).

Tallichet, Suzanne. "Women in the Mines: Stories of Life and Work," *NWSA Journal* 11, no. 3 (Fall 1999), 202–204.

Worthen, Helena. "Getting In: The Experience of Minority Graduates of the Building Bridges Project Pre-Apprenticeship Class," *Labor Studies Journal* 28, no. 1 (Spring 2003), 31–52.

MAGAZINE ARTICLES IN PRINT

Baird, Mary. "A Rich Treasure of Word Paintings Women's Lives in the World of Construction," *Tradeswomen*, April 30, 1998.

Bell, Dan. "Women Not Wanted: Female Construction Workers Face Chronic Unemployment and Daunting Odds. A New Mayoral Commission Will Have to Change the Face of an Industry," *City Limits*, May/June 2005, 12–15.

Bulger, Denise. "Electrician," *Tradeswomen Magazine*, Fall 1988, 57–58.

Cox, Clinton. "Fight Back . . . Goes After Construction Jobs," *Sunday News Magazine*, September 25, 1977.

Davis, Julie Bawden. "They're Up to the Task," *Los Angeles*, April 19, 1996.

Eisenberg, Susan. "Women Hard Hats Speak Out," *The Nation,* September 18, 1989.

Federation News. "Women in the Building Trades," May 2002.

Fischbach, Amy. "Wiring Women," *EC&M*, April 1, 2002.

Glamour Magazine. "Are your Tax Dollars Paying for Sexist Ed 101?" October 2002.

Martin, Molly. "Crossing the Country: Tradeswomen Do Washington," *Tradeswomen*, October 31, 1989.

Norberg-Johnson, Denise. "Are We There Yet?" *Electrical Contractor*, March 2004.

OCAW Reporter. "Tradeswomen's Conference Huge Success,' Say Organizers," November-December 1989.

Raskin, A. H. "No 25-Hour Week for Him," *New York Times Magazine*, March 18, 1962.

Sanders, Marion K. "James Haughton Wants 500,000 More Jobs," *New York Times Magazine*, September 14, 1969.

Techniques. "DOL Funds for Women in Apprenticeship and Nontraditional Occupations," January 2003.

The New York Enterprise Report. "Aurora Electric ranked as fastest-growing inner-city business," New York, July 1, 2002.

Van Gelder, Lindsy. "Blazing Battles," *Daily Magazine*, July 5, 1987.

Velie, Lester. "The Union That Gives More to the Boss," *Reader's Digest*, 1–5.

Watkins, Kristin. "Washington Notes," *Tradeswomen*, April 30, 1995.

Watson, Jennifer. "101st Congress Wrap-Up," *Tradeswomen*, April 30, 1990.

Whittemore, Hank. "'I' Want Respect That's All," *Parade Magazine*, July 10, 1988.

NEWSLETTERS

Allied Union News, November 28, 1941, June 22, 1945.

Fitch, Robert. "Wars of Attrition: Vietnam, the Business Roundtable, and the Decline of Construction Unions," *Hard Hat News*, Spring 2001.

In These Times. "Contractors make it tough for women," September 9–15, 1981.

RESIST Newsletter. "Still Alone in a Crowd," 10, no. 7, September 2001.

Union Labor Report. "Special Report Union Women Responding to Ongoing Sexism in Building Trades," 55, no. 30, July 26, 2001.

Work America Newsletter. "All-Craft Giving Women a Shot At Skilled Craft Jobs," 1, March 1984, Washington D.C.

NEWSPAPER ARTICLES IN PRINT

Associated Press. "Government Trying to Recruit Women for Trowel Trades," October 13, 1995.

———. "Judge Gives Preliminary OK to Ford Settlement in Civil Rights Complaint," February 11, 2005.

California Voice. "Women in Nontraditional Jobs," December 20, 1991.

Chicago Journal. "Chicago Women in Trades in the News," August 26, 2004.

Chicago Sun Times. "$2 Million Grant Beckons Women to Trades," October 5, 2004.

———. "Bill Would Give State Control of Union Apprentice Plans," April 14, 2005.

Chicago Tribune. "A Crummy Situation: Despite Workplace Gains, Women Still Aren't Making the Dough that Men Are," November 5, 2003.

———. "Apprentices' Wait is Usually Worthwhile," September 25, 2002.

———. " Grant Builds on Effort to get Women into Building Trades," October 5, 2004.

———. "Lack of College Shouldn't Keep Women Down," July 2, 2004.

Christian Science Monitor. "It's Dirty Work, and These Women Gotta Do It," January 6, 2003.

Deutsch, Claudia H. "Getting Women Down to the Site," *New York Times*, March 11, 1990.

Diario, Rodolfo C. "New York's day laborers want centers to come in from streets," New York Community Media Alliance, www.indypressny.org/nycma/voices/sources/Hoy/Rodolfo_Castillo_Diario/, July 29, 2005, Translated from Spanish by Hillary Hawkins.

Electrical Union World. "30 More Mfg. Division Members Finish Blue Print Reading Class," June 1971.

———. "300 Members of the Staten Island Electrical Club Attend Annual Dinner Dance," February 28, 1985.

———. "50 Apprentices Tryout for Softball Team," August 18, 1986.

———. "60ᵗʰ Anniversary of Welfare Club," June 30, 1983.

———. "80 'A' Apprentices Receive Community Service Certificates," February 1, 1984.

———. "'A' Division Steward Attend Meeting to Clarify Work Rule Changes," December 15, 1983.

———. "A" Negotiating Committee Selected at Wage & Policy Meeting."

———. "Allied Club Holds Annual 'Irish Fete,'" March 31, 1985.

———. "Allied Club Hosts Bayberry Land Seminar," November 11, 1986.

———. "Allied Club Marks 50ᵗʰ Anniversary," December 21, 1982.

———. "Apprentice Training Instructors Needed by Joint Apprenticeship Committee," October 15, 1985.

———. "Assembly Districts Played a Major Role," November 15, 1961.

———. "Bedsole Kingsboro Club: Goldin Adds to Ladies' Night," March 23, 1987.

———. "Bedsole Kingsboro Club Holds 54ᵗʰ Annual Dinner Dance," February 28, 1985.

———. "Brotherhood Party Is Not a 'Class Party' in Attack on Red Ideology," October 6, 1961.

———. "Brotherhood Party Rolls Along in High Gear," September 15, 1961.

———. "Business Manager Van Arsdale Installs Allied Club Officers," January 15, 1984.

———. "Catholic Council Celebrates 50ᵗʰ Annual Communion Breakfast,"

———. "Catholic Council Honors Pioneers," September 15, 1961.

———. "Catholic Council Hosts 26ᵗʰ Annual Ladies Night," April 21, 1988.

———. "Colony Club Holds Annual Beefsteak Party," January 12, 1987.

———. "Electchester, 'Greatest Project We've Seen'—Danish ECA Team," September 1951.

———. "Electchester May Be Completed By December," September 15, 1951.

———. "Electrical Welfare Club Marks 61 Years of Progress," November 15, 1984.

———. "Equal Opportunity Held 'Basic Remedy' for Birmingham," 1961.

———. "Farmed Inventor's Grandson Attends Latimer Dance," February 1, 1962.

———. "French ECA Team Visits Our Local, Electchester to Study U.S. Construction," June 1, 1951.

———. "Girls To Attend Camp Solidarity's 17th Camping Season," March 17, 1988.

———. "Gov. Dewey Says Local 3's Housing Project Should Be 'Model' for U.S.," September 15, 1949.

———. "How About Naming Your Housing Project-And Winning $100 U.S. Bond?" April 1, 1950.

———. "How Truman Fought for Housing," September 1, 1949.

———. "Jacob K. Javits Convention Center—New York's New 'Crystal Palace,'" May 1, 1986.

———. "Labor, Electrical Industry Leaders Hail Apprentice Training Instructors," February 10, 1970.

———. "Labor History Forming the International Brotherhood of Electrical Workers," Parts I, II, III, IV, V, VI, and VII, 1986.

———. "Ladies Night Held By Bedsole Club and Catholic Council," April 7, 1986.

———. "Latimer Group Makes First Scholarship Grant," October 1, 1961.

———. "Local 3 'A' Apprentice Applicants Begin Interviewing Process," December 15, 1983.

———. "Local 3 Gives $1,000 to Aid Reverend King," December 1, 1960.

———. "Local 3 Holds Puerto Rico Celebration," 1958.

———. "Local 3 Honors Scroll Ladies," 1975.

———. "Local 3 Officials Installed," July 15, 1950.

———. "Mayor Wagner Tells Shop Stewards Labor Has Vital Stake in City's Election," September 1, 1961.

——. "Member's Son Elevated to Auxiliary Bishop by Pope John Paul II," April 15, 1981.

——. "Members Tell of Ten-Day Visit to Virgin Islands and Jamaica," 1965.

——. "Members Vote $5 Weekly Savings Plan to Finance Electchester Construction," May 15, 1952.

——. "New Officers of the 'A' Apprentice Advisory Committee elected at Bayberry Seminar," January 15, 1984.

——. "New York City Cracks Down on Unlicensed Electrical Work," March 17, 1988.

——. "N.Y. City Labor Leaders Urge Shop Stewards Help Build Brotherhood Party," October 15, 1961.

——. "NYC's Construction Industry Found Deadlist," October 15, 1987.

——. "O'Dwyer Congratulates Us on Housing Project," October 18, 1949.

——. "Our Pioneer Members Were There," September 15 1961.

——. "Outtley Appointed To Committee On Exploitation," May 1, 1965.

——. "Phelps-Dodge Execs Visit Beautiful Electchester," December 1, 1952.

——. "Potofsky Honored By Urban League," July 15, 1955.

——. "President Tracy, Other IBEW Officials Visit Electchester," October 1, 1952.

——. "Protecting The Davis-Bacon Act," October 15, 1987.

——. "Recruiting New Members: Bedsole/Kingsboro Club," October 15, 1985.

——. "Rosenberg, Schmelzer Receive Electrical Welfare Club 1981 Service Awards," November 1, 1981.

——. "St. George Association Chapter 80 Officers Installed by Former President Blain," November 15, 1985.

——. "St. George Society: McSpedon-God and Country Award," January 19, 1989.

——. "St. George's 'God and Country Award,'" October 15, 1984.

——. "Staten Island Electrical Club Installs New Officers," March 20, 1986.

——. "Stork Visits Electchester First Time," June 1, 1952.

——. "Terence Cardinal Cooke Memorial Scholarship Established by Catholic Council of Electrical Workers," March 15, 1984.

——. "Thanks to People of Common Wealth," September 15, 1958.

——. "The French had a Word for Electchester," June 1, 1950.

——. "The Knights of Labor," July 15, 1984.

——. "The Testimony I am Hearing Convinces Me That We Need to Tighten the Law and Get it Enforced," March 23, 1983.

——. "Thomas Van Arsdale Installs New Bedsole Kingsboro Club Officers," July 15, 1981.

——. "Traveling Journeyman Attend Mets Game as Guests of Local 3," August 18, 1986.

——. "Truman Hopes Others Will Follow Our Home Project," August 8, 1949.

——. "Union-Busters Lose Big One," May 1, 1986.

——. "Van Arsdale, Other Union Leaders Pay Final Tribute to Reverend King at Funeral Services in Atlanta, Ga.," April 15, 1968.

——. "Van Arsdale Praises Rev. King for Leadership in Civil Rights Movement," April 1, 1964.

——. "Van Arsdale Receives Urban League Award," December 1, 1963.

——. "Van Arsdale Says: Our Brotherhood Party Is Here to Stay," November 16, 1961.

——. "Van Arsdale Speech on Brotherhood Party Program," October 6, 1961.

——. "White House Continues Attack on Davis-Bacon Act," May 1, 1986.

Finkel, Jerry. "Welfare Club Enters 65th Year," *Electrical Union World*, February 22, 1988.

————. "Welfare Club Holds Educational Seminar," *Electrical Union World*, August 27, 1987.

Genasci, Lisa. "Women Find Work in Traditionally Male Jobs But Inroads Being Made at Snail's Pace as Employees Meet Masculine Resistance," *Chicago Tribune*, March 20, 1995.

Goleman, Daniel. "Stereotypes of the Sexes Persisting in Therapy," *New York Times*, April 10, 1990.

Grant, Alison. "Making a Comparatively Good Wage," *Plain Dealer*, May 17, 2005.

Harris, Renee. "Profile in Courage: Local Women Chooses Nontraditional Career," *Miami Times*, November 10–16, 2004.

Hinds, Michael deCourcy. "Decision File," *New York Times*, May 7, 1982.

Jordan, Robert, "Women Gain Foothold in Trades Through Affirmative Action," *Boston Globe*, April 12, 1998.

Kelley, Christie Watts. "Construction women want greater roles in hardhats," *Business First*, March 5, 1999.

Kleiman, Carol. "Tradition-Bound Blue-Collar World Finally Sees its Barriers Brumbling," *Chicago Tribune*, June 15, 1987.

Kramer, Louis. "Doing Nails for a Living, With a Hammer," *New York Times*, June 12, 2005.

Lapham, James Sigurd. "Indians . . . To Farms . . . To Golf Courses . . . To Today This Was Electchester," *Electrical Union World*, October 10, 1975.

Lieberman, Ernest D. "Rebuild America by Rebuilding Labor," *New York Times*, January 8, 1989.

Lisser, Eleena De. "Women Gain Ground in Nontraditional Careers," *Asbury Park Press Community*, December 6, 1989.

Mann, Judy. "Job Discrimination Still a Menace," *Washington Post*, December 12, 1997.

Munro, Kevin. "Catholic Council Hosts 25th Anniversary Ladies Night," *Electrical Union World*. March 11, 1987.

————. "Catholic Council of Local 3: 27th Annual Ladies' Night," *Electrical Union World*. March 23, 1989.

New York Beacon. "Nontraditional Employment for Women Receives Grant," October 15, 1997.

New York Construction News. "Real Estate Meets Construction Women" September 2003.

New York Post. "Bedsole/Kingsboro Club: 30 Years of Scouting," May 12, 1962.

New York Times. "A New Housing Partnership," May 14, 1949.

————. "Group Says Course Training Still Breaks Along Sex Lines," June 7, 2002.

————. "Paul Tishman Gets Queens Housing Bid," September 16, 1950.

Preston, Julia. "5 Construction Executives are Accused of Paying Off Locals," *New York Times*, December 16, 2004.

Pugh, Thomas. "Brotherhood Hallmark at Electchester," *Sunday News*, May 25, 1975.

Racusin, M. Jay. "Employers Join with Labor in Housing Plan," *Electrical Union World*. May 13, 1949.

Riesel, Victor. "New Force on Political Scene," *New York Mirror*, September 7, 1961.

Robbins, Tom. "Plumbing the Depths," *Village Voice*, February 4–10, 2004.

————. "The Mob's Engineers," *Village Voice*, December 21, 2004.

————. "Rackets Remedy," *Village Voice*, December 14, 2005.

Rotella, Katie. "We Need More Female Plumbers," *Chicago Tribune*, October 24, 2004.

Sacramento Observer. "Jobs Directory for Women," November 13, 1991.

San Francisco Metro Reporter. "Women in Nontraditional Jobs," December 22, 1991.

Schierenbeck, Jack. "Pioneer Female Board Carpenter Fights for Job," *New York Teacher*, February 27, 2002.

Southwest News Herald. "More Outreach and Training Programs Ahead for Chicago Women in Trades," October 7, 2004.

Stahl, Gretchen. "Percentage of Workers in Nontraditional Jobs," *LA Times*, April 19, 1996.

The Star. "Women's Work: Pioneering Females are Making their Mark in the Trades," January 2, 2005.

Tribune. "Electchester Bicentennial Celebration, October 10, 1975.

Uchitelle, Louis. "For Blacks, a Dream in Decline," *New York Times*, January 15, 2007.

U.S. News and World Report. "Work of Their Own," February 2, 2003.

Vaughn, S. "Wanted: Women in Construction," *Los Angeles Times*, February 20, 1999.

———. "Wanted: Women in Construction-Industry Needs Workers, but Some Say Bias Still Remains," *New York Times*, February 20, 1999.

Washington Post. "Sex Bias Cited in Vocational Ed," June 6, 2002.

———. "Why Are All the Workmen . . . Men?" April 20, 2003.

Williams, Winston. "Port Authority 'Growing' Own Electricians," *New York Times*, January 3, 1989.

Zambito, Thomas. "Link Five Construction Firms to Union Payoffs," *Daily News*, December 16, 2004.

ONLINE JOURNAL AND NEWSPAPER ARTICLES

Baderschneider, Jean. "Women Changing Work," *Industrial and Labor Relations Review* 45, no. 4 (July 1992), http://links.jstor.org/sici?sici=0019-7939%28199207%2945%3A4%3C828%3AWCW%3D2.0CO%3B2-N.

Chen, David W. "Electchester Getting Less Electrical, Queens Co-op for Trade Workers Slowly Departs from Its Roots," *The New York Times*, March 15, 2004, http://query.nytimes.com/gst/fullpage/html, accessed May 29, 2007.

Epstein, Linda B. "What Is a Gender Norm and Why Should We Care? Implementing a New Theory in Sexual Harassment Law," *Stanford Law Review* 51, no. 1(1998), http://links.jstor.org/sici?sici=0038-9765%28199811%2951%3A1%3C161%3AWIAGNA%3E2.0CO%#B2-G.

Feagin Joe R. and Nikitah Imani. "Racial Barriers to African American Entrepreneurship: An Exploratory Study," *Social Problems* 41, no. 4 (1994), http://links.jstor.org/sici?sici=0037-7791%28199411%2941%3A4%3C562%3ARBTAAE%3E2.0CO%3B2-g.

"Is Pornography Bad," *The Ethical Spectacle* (1995), www.spectacle.org/1195/pornbad.html.

Kennedy, Genevieve M. and Douglas A. Wissoker. "An Analysis of the Correlates of Discrimination Facing Young Hispanic Job-Seekers," *The American Economic Review* 84, no. 3 (1994), http://links.jstor.org/sici?sici=0002822984%3A33C674%3AAAO TCO%3E2.0CO%3B2E.

Kerr, Daniel, and Christopher Dole. "Cracking the Temp Trap: Day Laborers Grievances and Strategies for Change in Cleveland, Ohio," *Labor Studies Journal* 29, no. 4 (2005), http://muse.jhu.edu/journals.labor_studies_journal/v029/29.4kerr.html.

Leonard, Jonathan S. "The Impact of Affirmative Action Regulation and Equal Employment Law on Black Employment," *Journal of Economic Perspectives* 4, no. 4 (1990), www.questia.com.

Lunn, John, and Huey L. Perry. "Justifying Affirmative Action: Highway Construction in Lousiana," *Industrial and Labor Relations Review* 46, no. 3 (1993), http://links.jstor.org/sici?sici=0019-7939%28199304%2946%3A3%3C464%3AJAAHCI%3E2.0CO%3B2-5.

"Mackinnon: Pornography is Oppression," *The Ethical Spectacle* (1995), www.specta-cle.org/1195/mack.html.

McLanahan, Sara S., and Erin, Kelly L. "The Feminization of Poverty: Past and Future," online article (1999), http://apps.olin.wustl.edu/macarthur/working%20papers/wp-mclanahan3.htm, accessed July 28, 2008.

Smith, Greg B. "Mob tax carved out of your $$," *New York Daily News*, September 9, 2007, www.nydailynews.com/front/story/349910p-298460c.html, accessed September 27, 2007.

Stombler, Mindy. "'Buddies' or "Slutties": The Collective Sexual Reputation of Fraternity Little Sisters," *Gender and Society* 8, no. 3 (1994), http://links.jstor.org/sici?siCi=0891-2432%28199409%298%3A3%22O%22TCS%3E2.0.CO%#b2-R.

OTHER ONLINE SOURCES

IBEW Jurisdictional Mapping System and Local Union Directory, www.ibew.org/directory, accessed July 24, 2007.

IBEW List of Products Made/Assembled and Services Provided by IBEW Members, www.ibew.org/products, accessed July 24, 2007.

National Joint Apprenticeship and Training Committee, www.njatc.org, accessed July 24, 2007.

Santiago Igleasias Club Educational Society: A Beginning, accessed July 24, 2007.

Wider Opportunities For Women, Inc. (WOW), Project Title: Six Strategies for Family Economic Self-Sufficiency, "Nontraditional Employment for Women," www.sixstrategies.org/sixstrategies/nontraditional.cfm, accessed July 26, 2007.

PAMPHLETS AND REPORTS

Agreement and Working Rules between Local Union No. 3 International Brotherhood of Electrical Workers, AFL-CIO and the New York City Chapter of the Electrical Contractors Association, May 10, 2007, http://local3ibew.org/?q=node/4369, accessed April 30, 2009.

"Apprenticeship Experience of Participants in New York State Apprenticeship Programs" (Labor Research Report Post No. 3), 1990–1991.

Association for Union Democracy, Women's Project, ""AUD Women's Project Launches 'Operation Punch List,' "www.uniondemocracy.org/UDR/8-operation%20punchlist.htm, October/November Issue of Union Democracy Review, accessed April 30, 2009.

Bailey, Thomas, and Roger Waldinger. "Labor Force Adjustments in a Growing Construction Market: The Metropolitan Area in the 1990s" (report prepared for the Planning and Development Department of Port Authority), 1987.

———. "The Increasing Significance of Race in the Construction Industry" (report prepared for the Port Authority), 1985.

Berik, Gunseli, and Cihan Bilginsoy. "Do Unions Help or Hinder Women in Training? Apprenticeship Programs in the U.S." Industrial Relations: Journal of Economy and Society, (The Regents of the University of California), Volume 39 Issue 4, December 17, 2002, 600–624.

Boyd, Mary Ellen. "Women in Non-Traditional Employment," *Fortieth Annual National Conference Proceedings on Labor*, New York University, 1990.

Broach, Howard. "Union Progress in New York: The Story of Modernization of Union Structure and Business Methods in the Electrical Field" in George Ruffini, *Harry Van Arsdale Jr., Labor's Champion* (New York: M.E. Sharpe), 2003, 253.

Davidson, H. Ron. "Harry Van Arsdale and the New York City Fiscal Crisis" (unpublished paper), August 1999.

Dinkins, David, and Dennis deLeon. *Building Barriers: A Report on Discrimination Against Women and People of Color in New York City's Construction Trades*, One Police Plaza, New York, 1993.

EEO-2 data on Joint Industry Board of the Electrical Industry, "Total Apprentices Accepted in JIB Apprenticeship Program. 1962–1972: A Breakdown According to Race and Gender," and "Non-White Participation in Apprenticeship in Selected Building Trades since March 1963" (compliance reports of the New York State Department of Labor).

Electrical Industry. "A Division Women's Conference at Bayberry Land" (educational seminar), February 28–March 1, 1992.

———. "A Portrait of Minority Participation in the Electrical Industry," in George Santiago, *Union Power and Affiliation*, (unpublished dissertation, 1987), 164, accessed at the Wagner Labor Archives, 1993.

———. "Club Presidents" (news reports), February 28, 1980.

———. "Wives, Husbands Teenagers and Breadwinners: How Unions Help Your Family" (publication of the Harry Van Arsdale Jr. Memorial Association), 198710, 11, 12, accessed at Empire State College (SUNY), HVA Library, 1992.

Grabelsky, Jeff. "Short Circuit: A History of the IBEW" (occasional papers series of the New York Labor History Association), 1991.

———. "Preserving Craft Pride: Autonomy and Control in the Construction Industry" (unpublished manuscript), 1990.

Hard Hatted Women. Funding proposal for security project to assist low-income and poor women into skilled construction jobs, Cleveland, Ohio, 2006.

Joint Apprenticeship Committee of the Joint Industry Board of the Electrical Industry. Affirmative Action Program, October 5, 1971.

Joint Apprenticeship Training Committee, New York District Council of Carpenters, United Brotherhood of Carpenters & Joiners of America. *Accepting Applications* (document), 2007.

Legal Defense and Education Fund. (Report and Recommendation of the NOW Legal Defense and Education Fund to the New York City Commission on Human Rights and the New York City Office of Labor Services on Discrimination in the Construction Trade), April 24, 1990.

Mayor Bloomberg's Commission On Construction Opportunities For Women, "10 Initiatives of Mayor's Commission on Construction Opportunity, New York City, January 15, 2007.

Negrey, Cynthia, Stacie Golin, Sunhwa Lee, Holly Mead, and Barbara Gault. "Working First But Working Poor: The Need for Education & Training Following Welfare Reform" (Executive Summary), September 2001.

New York City Commission on Human Rights. "Discrimination in New York City Construction Trades" (hearings), December 1993.

New York State Department of Labor, Division of Equal Opportunity Development. "Affirmative Action Report for the New York State Apprenticeship Council," July 16, 2001.

Office of the Secretary, Women's Bureau, U.S. Department of Labor. "Searching for a Job in the Construction Industry: Some Tips for Women," November 1979.

———. "Sources of Assistance for Recruiting Women for Apprenticeship: Programs and Skilled Nontraditional Blue-Collar Work," July 1978.

Marshall, Ray F., and Vernon M. Briggs Jr. "Negro Participation in Apprenticeship Programs," *Journal of Human Resources* 2, no. 1 (Winter, 1967): 51–69.

Santiago, George. *Union Power and Affiliation Within a Trade Union: Local #3 International Brotherhood of Electrical Workers* (Ph.D. dissertation printed in a pamphlet), New York: 1987.

Tradeswomen, Inc., "What Are Journey Sisters?" (pamphlet), San Francisco, CA, 1985.

U.S. Department of Labor. List of federally funded grants under the WANTO grant award to women's advocacy organizations providing pre-apprenticeship training into one skilled construction trade nationally (January 15, 2007), www.dol.gov/wb/03awards.htm.

———. "Women in the Construction Workplace: Providing Equitable Safety and Health Protection" (June 1999), http://www.osha.gov/doc/accsh/haswicformal.html.

U.S. Department of Labor, Bureau of Labor Statistics. "Twenty Leading Occupations of Employed Women," February 1990.

U.S. Department of Labor, Office of the Secretary, Women's Bureau. "The Women Offender Apprenticeship Program: From Inmate to Skilled Craft Worker", 1980.

———. "Women in Nontraditional Jobs: A Conference Guide: Increasing Job Options for Women", 1978.

———. "Facts on Working Women", January 1991.

———. "Women and Workforce 2000", January 1988.

Wider Opportunities for Women, Inc (WOW). "The Feminization of Poverty: An Update" (Publication of WOW), December 1983.

Women's Research and Education Institute (WREI) of the Congressional Caucus for Women's Issues. "Work & Women in the 1980s: A Perspective on Basic on Basic Trends Affecting Women's Jobs and Job Opportunities" (Publication of WREI), Washington DC, 1983.

PUBLIC DOCUMENTS

Agreement and Working Rules between NY Electrical Contractors Assoc., Inc., Assoc. of Electrical Contractors, Inc., and Local Union No. 3, IBEW, June 10, 1983.

Bylaws of Local Union No. 3, IBEW NY, February 14, 1985.

Local 3, Club Presidents as of February 2005.

Mayor Bloomberg, Press Release, *Objectives to Improve Access to Employment For Minorities, Women, Veterans and High School Graduates*, October 5, 2005.

New York City, Press Release, *Plumber Apprentices Wanted Now in New York City*, August 16, 2005.

The Women in Apprenticeship and Non Traditional Occupations (WANTO) Act, May 1996.

Union Labor Report, *Special Report: Apprenticeship Anxieties*, December 1, 1983.

U.S. Department of Labor, *A Woman's Guide to Apprenticeship*, Office of the Secretary, Women's Bureau, 1980.

———, *Employment in Perspective: Women in the Labor Force*, Bureau of Labor Statistics, 1988.

———, *Searching for a Job in the Construction Industry: Some Tips for Women*, Office of the Secretary, Women's Bureau.

———, *Sources of Assistance for Recruiting Women for Apprenticeship Programs and Skilled Nontraditional Blue-Collar Work*, Office of the Secretary, Women's Bureau, 1978.

———, *The Women Offender Apprenticeship Program: From Inmate to Skilled Craft Worker*, Office of the Secretary, Women's Bureau, 1980.

———, *Women in Nontraditional Employment: A Selected List of Publications, Slides and Films*, Office of the Secretary, Women's Bureau, 1979.

———, *Women in Non Traditional Jobs: A Conference Guide, Increasing Job Options for Women*, Office of the Secretary, Women's Bureau, 1978.

———, *Women in Non Traditional Jobs: A Program Model, Boston: Nontraditional Occupations Program for Women*, Office of the Secretary, Women's Bureau, 1978.

U.S. Department of Labor Current Population Survey, *Table 15: Median Usual Weekly Earnings of Full-time Wage and Salary Workers by Detailed Occupation and Sex, 2002 annual average*, Bureau of Labor Statistics, 2002.

Wider Opportunities for Women, *Statistics on Women and Nontraditional Work*, 2002.

SPEECH

Bond, Julian. "Hills to Climb and Strength to Climb Them," Speech given at Local Apprentice Graduation, New York, January 15, 1973.

Glossary

Italic boldface in text indicates glossary entries.

The American Business Council (ABC): The ABC has about 700 members in New York, including general contractors, subcontractors, insurance companies, and others involved in the construction industry. Of those, approximately 115 have state-approved apprenticeship programs. For recent developments on the activities of the ABC in New York, see www.bizjournals.com/albany/stories/2007/08/27/daily22 .html.

Apprenticeship Program: Consists of on–the-job training and classroom instruction. Apprentices must be eighteen years or older in order to apply for programs and programs must be approved by the state. The New York Apprenticeship Council is the regulating body that oversees apprenticeship programs in New York. The council reports to the New York State Department of Labor, an agency that is required by federal regulators at the Bureau of Apprenticeship and Training to file compliance reports for minority and women antidiscrimination laws.

Blue print: A type of copying method often used for architectural drawings. Blue prints are usually used to describe the drawing of a structure which is prepared by an architect or designer for the purposes of design, planning, estimating, securing permits, and actual construction.

Circuit breaker: A device that looks like a switch and is usually located inside the electrical breaker panel or circuit breaker box. It is designed to (1) shut off the power to portions of or the entire house and (2) to limit the amount of power flowing through a circuit (measured in amperes). 110 volt household circuits require a fuse or circuit breaker with a rating of 15 or a maximum of 20 amps. 220 volt circuits may be designed for higher amperage loads; for example, a hot water heater may be designed for a 30 amp load and would therefore need a 30 amp fuse or breaker.

Class "A" apprentice: Electrician trainee who is inducted into the five-year in-class training program and who received on-the-job training by *class "A" journeymen*. "A" apprentices are the only trainees that have a chance at obtaining a journeyman's "A" card.

Class "A" journeyman: First-grade electrician brought into the industry through the "A" *apprenticeship program*; most prestigious position in the industry and *craft union* with contractual obligation to work a five-hour day at straight time. Work on "A" rated jobs carry a great deal of overtime; sometimes resulting in journeymen earning nearly $80,000 per year. "A" journeymen are also referred to sometimes as "mechanics."

Class "M" journeyman: Second-grade electrician usually brought into the union through organizing a non-union shop. These are usually immigrant workers who know little or no English. "M" workers are assigned the least desirable journeymen work. "M" rated work is defined as *jobbing*, maintenance, or repair work. They are required to work an eight hour day at straight time. Overtime opportunities on "M" rated work is permitted without the permission of the business manager. "M" workers can make wages equivalent to "A" apprentices.

Construction Division A: The smallest yet most powerful decision-making segment of Local 3; comprised of construction "A" journeymen and "A" apprentices.

Construction Division M: This construction division is comprised of mainly immigrant workers, women, and *trainees*.

Contractor: A company licensed to perform certain types of construction activities. In most states, the *general contractor*'s license and some *specialty contractor*'s licenses don't require compliance with bonding, workmen's compensation, or similar regulations. Some of the specialty contractor licenses involve extensive training, testing, and/ or insurance requirements. There are various types of contractors.

Craft union: A union that admits only workers of a particular trade, skill set, or occupation (e.g., plumbers, carpenters, or electricians).

Deck jobs: New large nonresidential construction work sites where electricians work on open decks; that is, multi-level floors without walls. These jobs usually employ anywhere from 300 to 500 electricians and a few thousand workers from across a broad spectrum of the building trades. Apprentices complain that on deck jobs they are usually used for "gofer" work and that they rarely get an opportunity to learn the trade on these work sites. Women are always assigned to deck jobs because these projects are more likely to receive federal subsidies and be subject to compliance laws.

Divisions in Local 3, IBEW: There are over sixteen divisions in the union reflecting different occupational groups. The two largest are construction (8,000 members), which has different classes of craft workers, and maintenance, electrical manufacturing, and supply workers, who are spread out over 54 supply houses in the New York metropolitan area, as well as in New Jersey (18,000 members). Some of the divisions are:

- DBM Division
- EE Supply Division
- "F" Division

- Expeditors Division
- "A" Construction Apprentices
- "BACEE" Division – Stationary Engineers
- Electrical Inspectors Division
- OTB Technicians
- "M" Division
- "EE" Division – apprentice, trainee program
- "J" division – streetlighting and traffic signal group
- "J" division – yardman, painting, cleaning, storage, and garage dispatchers
- Newspaper and Tech Division
- White Plains "M" Division
- Sign Division
- ADM Division

Electrical conduit: A pipe, usually metal, in which wire is installed.

Electrical entrance package: The entry point of the electrical power, including: (1) the "strike" or location where the overhead or underground electrical lines connect to the house; (2) the meter that measures how much power is used; and (3) the "panel" or circuit breaker box (or fuse box), where the power can be shut off and where overload devices such a fuses or circuit breakers are located.

Electrical Workers Minority Caucus (EWMC): Founded in 1974 by the African American and Hispanic members of the IBEW who attended the 30th International Convention of the IBEW in Kansas City, Missouri.

Foreman: A union member who works as the leader of a construction work crew.

General contractor: Responsible for the execution, supervision, and overall coordination of a project and may also perform some of the individual construction tasks. Most general contractors are not licensed to perform all specialty trades and must hire *specialty contractors* for such tasks (e.g. electrical, plumbing).

General foreman: Directs the daily activities of the *shop foreman* and lead man and other construction employees in the preparation and completion of tasks.

Indenture: When an "*A" apprentice* passes all of the exams and interviews and has been assigned to a work site through a work ticket at the management-labor board.

Jobbing: Usually consists of renovation or maintenance building work that requires skilled craft work. It is usually handled by small building craft contractor firms in the building construction industry.

Joint Industry Board (JIB): Founded on March 30, 1943, when leaders from IBEW Local Union No. 3 and the electrical contracting industry recognized the need for an organization that would build and promote harmony within the electrical industry and address labor-management issues for electrical workers and contractors. Today, the JIB has developed into a renowned multi-employer organization which has become a role model for labor-management organizations throughout the state of New York and the country. The *JIB* administers member benefits (401(K), pension, medical, dental, etc.)

Local #3 Members: Local 3 Union Supply Division Shops have approximately 54 shops in the Bronx, Long Island, Queens, and New Jersey. They have a president and a shop steward for each manufacturing company.

Military Intermediate Journeyman (MIJ): A category that was devised by the *Joint Industry Board* to extend the terms of apprenticeship, which are usually from four to five years, an additional eighteen months. The MIJ receive lower wages than "A" class journeymen but higher than "A" class apprentices. An "A" apprentice will take an examination testing an apprentice's knowledge of electrical theory and electrical building codes in order to graduate to an MIJ. Unlike apprentices, MIJ can be drafted by the military. They have to take their "A" test in order to obtain a journeyman's card. Sometimes the MIJ test or the "A" test may be postponed for apprentices in order to extend their lower wage work in times of building downturns.

Open shop movement: Concerns the right of a worker not to be forced to join a union. In the construction industry, an association of small non-union contractors called the *American Business Council* (ABC) has become a formidable challenge to the unionized sector of the building trades unions.

Running work: Becoming a *foreman* and designing a work plan as well as supervising a work crew of usually five to ten workmen.

Shaping up: A practice dating back to the late nineteenth century whereby employers would notify dockworkers of jobs and ask them to assemble in front of ships at a specific dock. With regard to construction, workers are asked by employers to assemble on a specified street corner in the city where a truck picks them up for temporary day work. Currently, non-unionized contractors still use this practice mainly among immigrant and minority tradesmen.

Small shops: Usually renovation or maintenance work in offices. Personnel can number from three to twenty workers. Apprentices usually desire assignments to a small shop in order to have an opportunity to work more closely with a journeyman and learn a broad range of electrical trades skills.

Specialty contractor: Contractor licensed to perform a specialty task (e.g., electrical, side sewer, asbestos abatement, etc.).

Straw boss: A worker who is occasionally assigned to supervise a work crew on a construction site.

Subcontractor: A business in construction or an individual that signs a contract to perform part or all of the obligations of the *general contractor*.

Trainees: A group designated by the state to facilitate the entrance of minorities into construction brotherhoods who were excluded from the "A" apprentice programs. Trainees are resented by unionized construction workers and unionists as an intrusion of the state in their private business. Trainees usually have a longer route to be eligible to take the *MIJ* or "A" test. "A" apprentices can take their exams after five and one-half years, whereas trainees have to wait for eleven years. The union is not obligated to admit trainees into union membership. A subclass of apprentice—also considered substandard—who are admitted into the electrical program through either a state-sponsored program required by antidiscrimination legislation or an organizing drive by the Local on construction sites whereby non-union workers sign union cards.

Tramp: A traveler from another electrician local who is work in the Local 3's jurisdiction.

United Electrical Contractors Association (UCEA): An employer/owners association of building companies that negotiates with Local 3's leadership for collective bargaining agreements. The UECA tried to seek legal recourse to avoid negotiating with the Local in 1998, and only returned to the bargaining table in May 2008.

Work crew: A team of electricians, ranging anywhere from three to ten workers. Work crews are composed of journeymen and apprentices who are assigned to a specified area of the work site to carry out a work task under the supervision of a temporary foreman, called a *straw boss*, who, in turn, is supervised by a regular *foreman*, who is supervised by a *general foreman*, usually with an assistant general foreman, all of whom are in the union.

Work rules: Contractual agreements (usually numbering approximately ten) that outline the behavior and jurisdiction of workmen vis-à-vis employers. The most important work rules pertain to the ratio of apprentices to journeymen permitted on certain types of construction jobs. They are as follows:

(1) One apprentice to each journeyman on all commercial, industrial, and residential jobs up to $10,000.
(2) One apprentice to each journeyman on all one, two, three, and four family houses.
(3) One apprentice to three journeymen or fraction thereof on new construction and alteration work. This ratio is interpreted to allow the following apprentice to journeyman relation on any job or in any shop:

1 Apprentice to 1 Journeyman
1 Apprentice to 2 Journeymen
1 Apprentice to 3 Journeymen
2 Apprentices to 4 Journeymen
2 Apprentices to 5 Journeymen
2 Apprentices to 6 Journeymen
3 Apprentices to 7 Journeymen

Work ticket: Upon layoffs for jobs completed, tradesmen return to the management-labor board hiring hall to request a job assignment. These "tickets" are issued at the window of the office.

Index

Francine A. Moccio is Director of the Institute for Women and Work, ILR School at Cornell University.